Environmental Health: Protection and Sustainability

Environmental Health: Protection and Sustainability

Editor: Millie Donovan

Callisto Reference

www.callistoreference.com

Callisto Reference,
118-35 Queens Blvd., Suite 400,
Forest Hills, NY 11375, USA

Visit us on the World Wide Web at:
www.callistoreference.com

ISBN: 978-1-64116-223-4 (Hardback)

Cataloging-in-Publication Data

Environmental health : protection and sustainability / edited by Millie Donovan.
p. cm.
Includes bibliographical references and index.
ISBN 978-1-64116-223-4
1. Environmental health. 2. Environmental protection. 3. Sustainability.
4. Sustainable development. I. Donovan, Millie.
RA565 .E58 2019
363.7--dc23

Table of Contents

Preface

This book has been an outcome of determined endeavour from a group of educationists in the field. The primary objective was to involve a broad spectrum of professionals from diverse cultural background involved in the field for developing new researches. The book not only targets students but also scholars pursuing higher research for further enhancement of the theoretical and practical applications of the subject.

Environmental health is a branch of public health, which studies all aspects of the natural and built environment that affect human health. Five significant disciplines integral to the field of environmental health include toxicology, environmental epidemiology, environmental engineering, exposure science and environmental law. Toxicology, epidemiology and exposure science studies are integrated to conduct a risk assessment of chemicals, specific chemical mixtures and health risk factors. These studies are then used for regulating chemical emissions, developing and implementing environmental health policies such as imposing standards for proper sanitation, etc. Some of the areas of focus in environmental health include indoor and outdoor air quality, climate change, noise pollution control, liquid waste disposal, solid waste management, toxic chemical exposure, vector control, medical waste disposal and management, land use planning, etc. This book elucidates the concepts and innovative models around prospective developments with respect to environmental health. It presents researches and studies performed by experts across the globe in relation to the protection and sustainability of the environment. It attempts to assist those with a goal of delving into this field.

It was an honour to edit such a profound book and also a challenging task to compile and examine all the relevant data for accuracy and originality. I wish to acknowledge the efforts of the contributors for submitting such brilliant and diverse chapters in the field and for endlessly working for the completion of the book. Last, but not the least; I thank my family for being a constant source of support in all my research endeavours.

Editor

1

Energy-Saving Scheduling in a Flexible Flow Shop using a Hybrid Genetic Algorithm

Rong-Hwa Huang[1], Shun-Chi Yu[2], Po-Han Chen[1]

[1]Department of Business Administration, Fu Jen Catholic University, Taiwan
[2]Department of Tourism and Hospitality, China University of Science and Technology, Taiwan
Email: 026299@mail.fju.edu.tw

Abstract

Many researches discussing reduced energy consumption for environmental protection focus on machine efficiency or process redesign. To optimize the machine operation time can also save the energy, and these researches have received great interests in recent years. This study considers three different states of machines, among processing there are two different speeds, to solve the problem of minimizing energy costs under time-of-use tariff with no tardy jobs in flexible flow shop. This problem is basically NP-hard, we proposed a hybrid genetic algorithm (GA) to solve problems in reasonable timeliness. The result shows that to optimize different states of machines under time-of use tariff can reduce energy costs significantly in on-time delivery.

Keywords

Flexible Flow Shops, Energy-Saving Genetic Algorithm, Energy Consumption Cost, Non-Tardy, Genetic Algorithms

1. Introduction

Goldratt and Cox [1] posit that goal achievement by a goal-oriented system is limited by at least one constraint. The limitation of electricity affects the energy in Taiwan. Good energy-saving production systems have become important issue during peak time. In practical production, energy saving is essential for enhancing production activity and maximizing the effectiveness of processes on machines. The electricity consumption of industries exceeds half of global consumption and affects the operating efficiency in practical situations.

A flexible flow shop (FFS) scheduling problem provides multiple identical parallel machines at each station for increasing capacity and reducing costs.

Flow shop processing involves constant sequences of initial, standard, steady, and nonpermutable production, such as that of steel, optoelectronics, and metal, which are common in real-world situations [2] [3]. To offer widely application, Linn and Zhang [4] indicated that FFS scheduling applies to traditional flow shop situations. The attempted problem is excellently solved. However, the energy consumption of machines affects the efficiency and quality of production; for instance, increased on-peak electricity consumption of machines causes an increase in temperature, wearing of parts, and improper manual operation. The time-of-use (TOU) tariff is a common used for adjusting peak power consumption worldwide. Generally, the billing period is divided into on-peak hours, mid-peak hours, and off-peak hours. The electricity costs during the on-peak period are the highest, whereas those during the off-peak period are the lowest. Moreover, machines can be switched to operation, standby, and shutdown modes for preventing unnecessary energy wastage.

As for trade-off of energy saving method, Ribas *et al.* [5] demonstrated that enterprises added machines to certain stages, thus enhanced production efficiency and customer satisfaction. An energy-efficient mathematical model for solving FFS scheduling problems and multiobjective optimization to minimize the makespan and total energy consumption [6]. The experimental results showed that the relationship between the makespan and energy consumption may be conflicting. A proposed energy-saving decision is useful for minimizing energy consumption. To minimize trade-off of earliness and tardiness in a practical production environment, Huang *et al.* [7] developed a farness particle swarm optimization (FPSO) algorithm for solving reentrant two-stage multiprocessor flow shop scheduling problems.

As for energy-saving issues, because of scarce resources, increasing attention has been focused on developing practical production techniques for clean technologies and protecting the environment [8]. Decreasing emission of CO_2 attracts more attentions. Zhang *et al.* [9] developed a time-indexed integer programming formulation for solving manufacturing schedules that minimize electricity cost and the carbon footprint under TOU tariffs without compromising production throughput. The proposed method used a flow shop with eight process steps that were operated on a typical summer day for an energy-saving test. Results indicated that shifting electricity usage from on-peak hours to mid-peak hours or off-peak hours reduces electricity costs; however, it may increase CO_2 emissions in regions where the grid base load is met with electricity from coal-fired power plants. The tradeoff between minimizing electricity cost and reducing CO_2 emissions was shown using a Pareto frontier. Setlhaolo [10] demonstrated that using residential demand response along with a mixed integer nonlinear optimization model under a TOU electricity tariff can solve the scheduling problem of typical home appliances as well as minimize electricity costs and facilitate earning relevant incentives [11]. An analysis revealed that households can alter their electricity consumption in response to varying prices and incentives; thus, a consumer may reduce electricity costs by more than 25%. Different values of the weighting factor (α) provide varying costs. According to the α values and their preferences, consumers can choose their electricity costs. FFS scheduling can save energy and lower manufacturing costs by clean technologies, environmental policy, reducing the completion time and inventory in industrial environments.

Energy-saving models are created for important issues, Mouzon and Yildirim [12] developed operational methods for minimizing the energy consumption of manufacturing equipment and total completion time. The energy used in processes can be saved by turning off nonbottleneck (*i.e.*, underutilized) machines or equipment when they are idle for a certain period of time. In particular, an analysis indicated that the proposed dispatching methods are effective in reducing the energy consumption of underutilized manufacturing equipment. Therefore, a production manager has a set of nondominated solutions (*i.e.*, a set of efficient solutions) that he or she can use for determining the most efficient production sequence; moreover, they minimize the total energy consumption while optimizing the total completion time. Fang *et al.* [13] developed a new mathematical programming model for flow shop scheduling problems to minimize the peak power load, energy consumption, and associated carbon footprint along with the cycle time. In a flow shop with two machines producing various parts, the operation speed was considered an independent variable, which can be changed to affect the peak load and energy consumption. The results demonstrated that the proposed approach enables determining near-optimal schedules for achieving energy-saving goals. Tibi and Arman [14] developed a mathematical linear programming model to optimize the decision-making for managing a cogeneration facility as a potential clean-development mechanism project in a hospital in Palestine. The model optimized the cost of energy and the cost of installation of a small cogeneration plant under constraints on electricity-and-heat supply and demand balances. The results proved the efficiency of the proposed method. He *et al.* [15] demonstrated that the environmental load resulting from the energy consumption of machine tool systems is broadly recognized. Improving scheduling saves energy, facilitates efficient use of machine tools, and reduces energy consumption by idle equipment. One proposed energy-saving optimization method involves machine tool selection and a series of machine operations for flexible job shops. The method was designed to reduce the energy consumption of machine operations, and the scheduling was aimed at reducing the unused power consumption of machine tools. The current study investigated how to develop and use clean technologies like non-tardy procedures for scheduling the use of parallel machines to maintain practical production for environmental protection.

To prevent tardiness and ensure effective operation of production equipment, companies create effective and efficient production environments that maximize corporate benefits from production activities. Bruzzone *et al.* [16] developed energy-aware scheduling of manufacturing processes by using advanced planning and scheduling, a mathematical model for optimally planning energy saving for a given schedule. The new approach relies on the MIP model, where the reference schedule is modified to account for energy consumption without changing the jobs' assignments and sequencing provided by the reference schedule. The results demonstrated that a commercial MIP solver and an original MIP heuristic are applicable in practical production.

Enumeration and heuristic methods have been applied for energy saving in previous studies [10]. Integer programming, branch and bound programming, and MIP are the most widely used enumeration methods that can provide appropriate solutions. However, high computational times limit the applicability of enumeration methods to small-scale problems [17]. Thus, heuristics such as the genetic algorithm, simulated annealing algorithm, and ant colony optimization algorithm are commonly used for solving energy-saving problems. Lian [18] obtained the average relative error rates of −28.20% and 60.25% for a combined local and global PSO algorithm against PSO and genetic algorithms, respectively. Zhang *et al.* [19] presented an I-ATTPSO algorithm with an average effectiveness improvement rate of −14% in small-scale problems and 55% in large-scale problems. Liu *et al.* [20] obtained an average relative error rate of 0.65% for the PSO-EDA_PI algorithm against other algorithms. Zhao *et al.* [21] found the average relative error rate of their proposed logistic dynamic PSO algorithm against other algorithms to be approximately 1.19% - 2.39%.

In many industrialized countries, manufacturing industries pay stratified electricity charges depending on the time of the day (*i.e.*, on-peak, mid-peak, and off-peak hours) [22]. China saves energy concerning the impact of internal industrial configuration in terms of size and ownership structure on aggregate energy intensity [23]. Besides, Germany property owners can deduce where they should ideally invest in order to optimize the energy efficiency of their building stock sustainably [24]. By contrast, the emerging smart grid concept may demand that industries pay real-time hourly electricity costs for the efficient usage of energy. To enable decision makers to apply feasible solutions for resolving unrelated parallel machine scheduling problems, Moon *et al.* [25] developed an energy-efficient method by using the weighted sum objective of production scheduling and electricity usage. Reliability models using a hybrid genetic algorithm along with their blank job insertion algorithm consider the energy cost aspect of the problem with the objective function of optimizing the weighted sum of two criteria: the minimization of the production makespan and the minimization of time-dependent electricity costs. The results demonstrated its performance in simulation experiments in practical production.

Energy-saving method relates to fixed costs. Shrouf *et al.* [26] posited that rising energy costs associated with increased production costs at manufacturing facilities encouraged decision makers to tackle this problem in different ways. One crucial step in this trend is to reduce the energy consumption costs of production systems. Considering variable energy prices in a single day, the authors proposed a mathematical model for minimizing energy consumption costs for single machine production scheduling during production processes. The job processing time consists of the starting time, idle time, and times when the machine must be shut down, turned on, and turned off. The proposed mathematical model enables the operation manager to implement the least expensive production scheduling during a production shift. To obtain feasible solutions by using a genetic algorithm, this study also determined whether the heuristic solution provides the minimum cost and optimal schedule for minimizing energy costs. In addition, an analytical solution was applied to generate the optimal solution. Moreover, analytical solutions and heuristic solutions were compared, and the heuristic solution is considered preferable for larger problems. The results indicate that significant reductions in energy costs can be achieved by avoiding high-energy price periods. The results have a positive environmental effect by reducing energy consumption during peak periods, thereby increasing the possibility of reducing CO_2 emissions from power generator sites. Although a genetic algorithm can efficiently solve energy-saving problems and prevent entrapment at a local optimum, the current study improves the solving process for enhancing performance in practical production and environmental protection.

By applying the nontardy constraint to practical production, the current study attempted to minimize makespan costs. A Cmax minimization genetic algorithm (CGA) and energy minimization genetic algorithm (EGA) are proposed for use in the first stage of solving two-stage multiprocessor flow shop scheduling problems and minimizing makespan costs. Furthermore, an adjusted Cmax minimization genetic algorithm (ACGA) and adjusted energy minimization genetic algorithm (AEGA) used in the second stage were compared for obtaining superior solutions and aiding enterprises in increasing profits and lowering overhead costs. The current study compared the proposed solution with two reported solutions that yielded comparable improvements.

The remainder of the paper is organized as follows: In Section 2, the FFS is formulated. In Section 3, the basic algorithms are introduced briefly. Then, the framework of energy-saving genetic algorithm for solving the FFS is proposed in Section 4. The influence of parameter setting is investigated based on design of experiment testing in Section 4, and computational results and comparisons are provided as well. Finally, we end the paper with some conclusions in Section 5.

2. Problem Definition

According to notation rule of Pinedo [27], the current study formulates the problem as $FF_k \mid EP_t, State \mid EC \mid T_j = 0$. FF_k represents k-stage flexible flow shop scheduling environment. EP_t means different electricity bill in each period. $State$ denotes three conditions of operation: stand-by, or shut-down of machines, and re-operation of machines also consumes power. $EC \mid T_j = 0$ demonstrates all operations of non-tardy jobs must complete before completion dates in order to minimize electricity costs. **Figure 1** demonstrates the Gantt

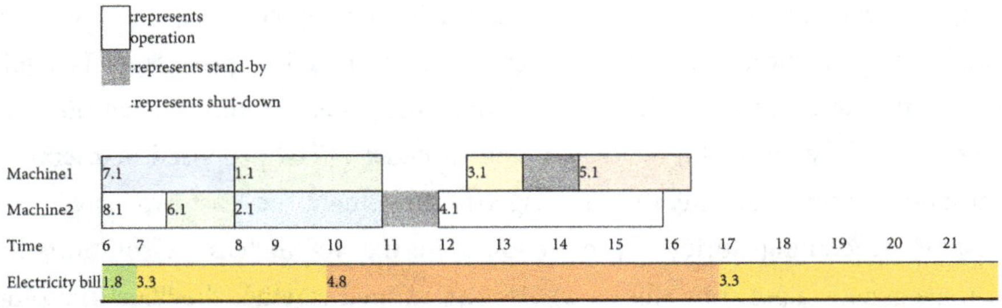

Figure 1. State of machine process.

chart of the attempted problem.

The FFS problem provides multiple identical parallel machines at each station for increasing capacity and reducing costs in practical production. One machine at every station can be selected for each job and the process begins from the first station. After the final job is completed in the second stage, all jobs are completed.

The scope and constraints of this study are demonstrated as follows:

1) Number of jobs, machines and stages are known.

2) Processing time of each job on each machine is known and constant.

3) Each machine in each stage can only process one job simultaneously.

4) The sequence through which jobs pass through machines may differ with the sequence of machine receiving job.

5) Jobs cannot be preempted.

6) There is no permutation, or machine breakdown.

7) The job ready time is 0.

8) Machines can switch into operation, stand-by, and shut-down.

2.1. Notation

T = total period set, all jobs in this study

M = total parallel machines at stage i, all machines in this study

D = due date set, all jobs completed dates in this study

J = job set, all jobs waiting for scheduling in this study

K = stage set, all stages in this study

$C_{\max}$ = the maximal completion time of all jobs

d_j = the due date of job J_j

C_{jkm} = the completion time of job J_j processed in the m-th machine at the k-th stage

S_{jkm} = the starting time of job J_j processed in the m-th machine at the k-th stage

P_{jkm} = the processing time of job J_j processed in the m-th machine at the k-th stage

E_t^O = the turn-on energy consumption of all machines at t-th time

E_t^I = the stand-by energy consumption of all machines at t-th time

E_t^R = the working energy consumption of all machines at t-th time

e_{km}^O = the working energy consumption in the m-th machine at the k-th stage

e^{I}_{km} = the stand-by energy consumption in the m-th machine at the k-th stage

e^{R}_{jkm} = the working energy consumption of job J_j processed in the m-th machine at the k-th stage

EP_t = the electricity bill of different time period

α_{jkmt} = indicator of whether job J_j is scheduled at the k-th stage in m-th machine at t-th time (α_{jkmt} = (0, 1); if α_{jkmt} = 1, job J_j is processed at the k-th stage in m-th machine at t-th time; otherwise α_{jkmt} = 0)

β_{jkmt} = indicator of whether job J_j is assigned at the k-th stage in m-th machine at t-th time (α_{jkmt} = (0, 1); if α_{jkmt} = 0, job J_j is processed at the k-th stage in m-th machine at t-th time; otherwise α_{jkmt} = 1)

Y_{kmt} = indicator of whether at the k-th stage the m-th machine turned on at t-th time (Y_{kmt} = (0, 1); if Y_{kmt} = 1, the m-th machine was turned on at t-th time; otherwise Y_{kmt} = 0)

δ_{kmt} = indicator of whether at the k-th stage the m-th machine turned off at t-th time (δ_{kmt} = (0, 1); if δ_{kmt} = 0, the m-th machine was turned off at t-th time; otherwise δ_{kmt} = 1)

2.2. Mathematical Model

1) Objective function

$$\min \sum_{t\in T} EP_t\left(E^R_t + E^I_t + E^O_t\right) \tag{1}$$

Equation (1) is the objective formulation and is primarily designed for minimizing energy consumption costs, such as those during operation, standby mode, and working. This equation measures the common criterion of the completion of all jobs and aids enterprises in improving the energy consumption and efficiency of production scheduling.

2) Total energy consumption

$$E^I_t = \sum_{j\in J}\sum_{k\in K}\sum_{m\in M} \beta_{jkmt}\delta_{kmt}e^I_{km},\ t\in T \tag{2}$$

$$E^R_t = \sum_{j\in J}\sum_{k\in K}\sum_{m\in M} \alpha_{jkmt}e^R_{jkm},\ t\in T \tag{3}$$

$$E^O_t = \sum_{m\in M} \gamma_{kmt}e^O_{km} \tag{4}$$

Equation (2) is the standby total energy consumption of all machines at the t-th time. Equation (3) is the total energy consumption of all machines in the t-th time period. Equation (4) is the turn-on total energy consumption of all machines in the t-th time period.

3) Completion time of job

$$C_{jkm} \le d_j, j\in J, k\in K,\ m\in M \tag{5}$$

$$C_{jkm} = S_{jkm} + P_{jkm}, j\in J, k\in K,\ m\in M,\ t\in T \tag{6}$$

$$C_{jkm} \le S_{(j+1)km}, j\in J, k\in K,\ m\in M \tag{7}$$

$$C_{jkm} \le S_{(k+1)m},\ j\in J, k\in K,\ m\in M \tag{8}$$

Equation (5) is the time for completing job j before the due date. Equation (6) is the completion time of job *j* by the *m*-th machine at the *k-th* stage and equals the starting time plus the processing time. Equation (7) is the completion time of job *j* by the *m*-th machine at the *k-th* stage and is no longer than the starting time of job (*j* + 1) for the *m*-th machine at the *k-th* stage. Equation (8) is the completion time of job *j* by the *m*-th machine at the *k-th* stage and is no longer than the starting time of job *j* for the *m*-th machine at the (*k* + 1)-*th* stage. Because of continuous production, the processing time prolongs the completion time and consumes energy. Therefore, minimizing energy consumption can remain the primary objective of machines for further production.

4) The constraints of job processing

$$\sum_{m\in M} \alpha_{jkmt} = 1, j = J,\ k \in K,\ t \in T \tag{9}$$

$$\gamma_{kmt} + \delta_{kmt} \leq 1,\ k \in K,\ m \in M,\ t \in T \tag{10}$$

$$\alpha_{jkmt} = \delta_{kmt},\ j \in J,\ k \in K,\ m \in M,\ t \in T \tag{11}$$

Equation (9) controls all jobs during each stage and ensures that they are processed only once. Furthermore, Equation (10) controls the *m*-th machine and ensures that only one job can be processed by it. Finally, Equation (11) determines whether the *m*-th machine is turned off at the *t*-th time (δ_{kmt} = (0, 1); if δ_{kmt} = 0, the *m*-th machine was turned off at *t*-th time; otherwise δ_{kmt} = 1.

3. Concept of Energy-Saving Genetic Algorithm Solving Procedure

According to $FF_k \mid EP_t,\ State \mid EC \mid T_j = 0$, this study constructed energy-saving genetic algorithm (ESGA) using two-stage solving procedures: a genetic algorithm for maximizing the makespan and another for minimizing energy consumption in the second stage. **Figure 2** demonstrates the starting time of solutions applied in the first stage to avoid on-peak hours.

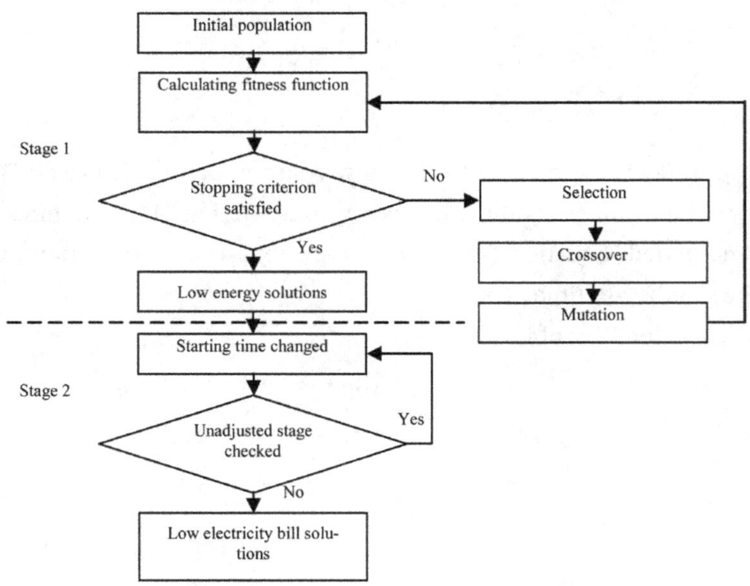

Figure 2. Flow chart of solving procedure of ESGA.

3.1. The First Stage of ESGA

This study utilized two genetic algorithms for solving problems: the CGA for minimizing the makespan and calculating the electricity cost and EGA for determining low energy consumption scheduling and calculating the electricity cost concurrently [6]. The procedures of the proposed method are demonstrated as follows:

3.1.1. Encoding

According to Dai *et al.* [6], a job set, stage set, and machine set ($s = 1, 2, \cdots, k$) are present at each stage. This study formulated FFS scheduling as a genetic matrix in which the columns are jobs and stages are rows.

$$A_{k\times j} = \begin{bmatrix} a_{1,1} & \cdots & a_{1,j} \\ \vdots & a_{1,1} & \vdots \\ a_{k,1} & \cdots & a_{k,j} \end{bmatrix} \tag{12}$$

The $a_{m,n}\left(m = 1, 2, \cdots, k, n = 1, 2, \cdots, j\right)$ is a real number in the interval (1, M_s). $a_{m,n}$ indicates which machine processes job *j*, and the decimal indicates the processing sequence; the lower the decimal is, the earlier the job is processed. A coding matrix represents a chromosome in which $k \times j$ genes are present.

$$\left[a_{1,1}, a_{1,2}, \cdots, a_{1,j}, a_{2,1}, a_{2,2}, \cdots, a_{k,j}\right] \tag{13}$$

For example, in FFS scheduling, the coding matrix consists of three stages, in which there are two machines and eight jobs to be processed.

$$A = \begin{bmatrix} 1.301 & 2.533 & 1.415 & 2.76 & 1.824 & 2.351 & 1.113 & 2.204 \\ 1.261 & 1.997 & 2.442 & 1.528 & 2.609 & 1.016 & 2.224 & 2.185 \\ 1.518 & 1.635 & 2.254 & 2.965 & 2.753 & 1.378 & 2.181 & 2.156 \end{bmatrix} \tag{14}$$

Row 1 represents processing conditions of the first stage. The $a_{1,1} = 1.301$ means J_1 processed in machine 1, $a_{1,2} = 2.533$ means J_2 processed in machine 2, and $a_{1,3} = 1.145$ means J_3 processed in machine 1. The decimal of $a_{1,1}$ is 0.301smaller than 0.415 of $a_{1,3}$, which represents $a_{1,1}$ is processed earlier than $a_{1,3}$.

$$\begin{aligned}&[1.301, 2.533, 1.415, 2.76, 1.824, 2.351, 1.113, 2.204, 1.261,\\ &1.997, 2.442, 1.528, 2.609, 1.016, 2.224, 2.185, 1.518, 1.635,\\ &2.254, 2.965, 2.753, 1.378, 2.181, 2.156]\end{aligned} \tag{15}$$

3.1.2. Initial Population

The initial solution is the first population of a genetic algorithm. Generally, two methods can generate initial solutions: one is randomly generated and other requires research. The scale of the initial population affects the efficiency and quality of solutions. A random initial solution was used in this study.

3.1.3. The Fitness Function

The fitness function can identify the quality of chromosomes, enabling inferior solutions to be screened out. The fitness function of the CGA is as follows:

$$f(x)=\frac{1}{C_{\max}} \tag{16}$$

The fitness function of EGA is as below:

$$f(x)=\frac{1}{\sum_{t\in T} EP_t\left(E_t^T+E_t^I+E_t^O\right)} \tag{17}$$

3.1.4. Selection

The algorithm selects a favorable parent from chromosomes and proceeds with crossover and mutation. Each chromosome has a fitness value, which determines the possibility for crossover and mutation. This study adopted roulette wheel selection.

3.1.5. Crossover

The proposed algorithm partially exchanges two chromosomes. In this study, the exchange rate was set as, which is a rounded-up approximation of the quantity of exchanged chromosome multiplied by the exchange rate. If the quantity of exchanged chromosome is odd, then one chromosome is added to the total chromosomes and the quantity becomes even. The exchange rate was set at 0.9, and this study adopted a two-point crossover method to solve the considered problem.

3.1.6. Crossover

Mutation can generate multiple and various children. A mutation rate of P_m was set in this study. The proposed method arbitrarily generates a probability value. If the value is below the mutation rate, mutation occurs. The mutation rate was set at 0.1. The mutation rate is uniformly distributed from 0 to 1. If the rate becomes 1, the proposed algorithm regenerates an interval of real numbers (1, M_s + 1).

3.2. The Second Stage of ESGA

Although the genetic algorithm generates scheduling with lower electricity costs, some jobs still generate high electricity costs. The solutions generated from the GCA and EGA in the first stage were adjusted using the ACGA, and problems in the second stage were solved using the AEGA to avoid high-energy price periods.

The adjustment rule begins from the final stage and proceeds in descending order. The proposed approach organizes the operations into the final order. The method gradually schedules operations in the off-peak period and moves the sequence of operation away from the on-peak period until all jobs are scheduled. **Figure 3** illustrates the flow chart.

3.3. A Brief Example

In a job scheduling scheme consisting of three stages, two machines at each stage, and eight jobs to be processed, the total working time is 16 hours and the factory working period is from 6 a.m. to 10 p.m. on each day. **Table 1** shows the operation time:

Table 2 shows the consumed energy of machines. The starting time of the machines is assumed to be short with sudden turn-on energy consumption.

Table 3 shows the electricity costs of every time period. The parameters of the proposed method are as follows: initial population, 30; exchange rate, 0.9; mutation rate, 0.1; and stopping criterion, 300 iterations.

3.3.1. The First Stage of ESGA

Figure 4 shows the computational result using CGA, and the electricity bill is NT$ 3277.8.

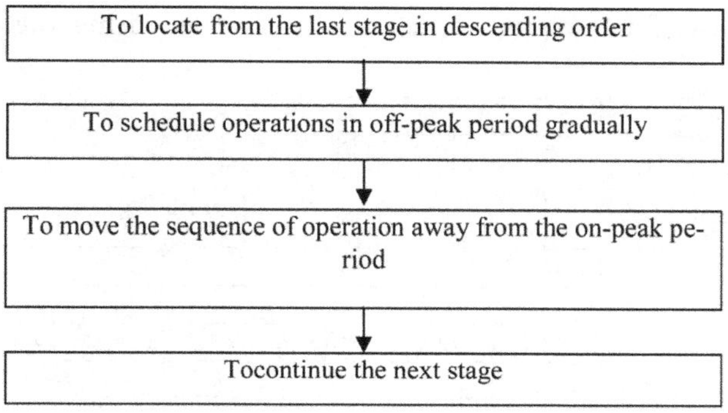

Figure 3. Flow chart of solving procedure of ESGA.

Table 1. The operation working time.

Jobs		The working time of jobs at different stages		
		Stage 1	Stage 2	Stage 3
1	Hour	3	4	1
2	Hour	3	2	2
3	Hour	1	2	3
4	Hour	4	2	3
5	Hour	2	2	0.5
6	Hour	0.5	1	1.5
7	Hour	2	0.5	1
8	Hour	1.5	2	1.5

Table 2. The operation working time.

Conditions of machines	Stage 1	Stage 2	Stage 3
Stand-by (kW)	3	7	6
Operation (kW)	8	18	17
Turn-on (kW)	4	10	8

Remarks: This study assumes the starting time of machine is short with a sudden turn-on energy consumption.

Figure 5 shows the computational result using EGA, and the electricity bill is NT$ 2900.6.

3.3.2. The Second Stage of ESGA

The adjustment rule begins from the final stage and proceeds in descending order. The proposed approach organizes operations into the final order. The method gradually schedules operations in the off-peak period and moves the sequence of operation away from the on-peak period until all jobs are scheduled. **Figure 6** shows the computational result obtained using the ACGA and the electricity cost is NT$ 2929.7.

Figure 7 the computational result using AEGA, and the electricity bill is NT$ 2807.4.

Table 4 lists four computational results obtained using the CGA as the

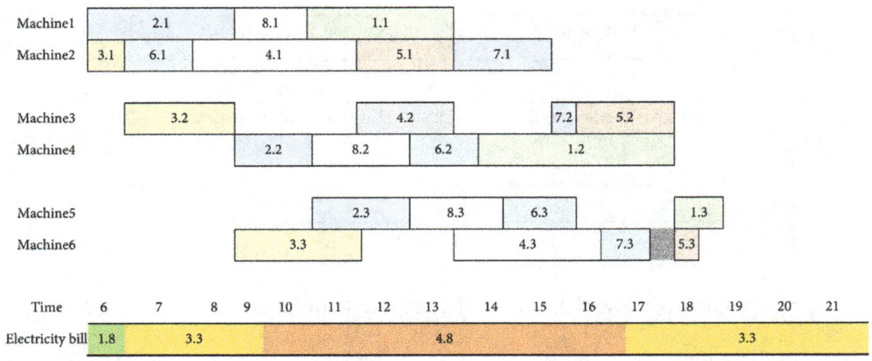

Figure 4. The Gantt chart of computational result using CGA.

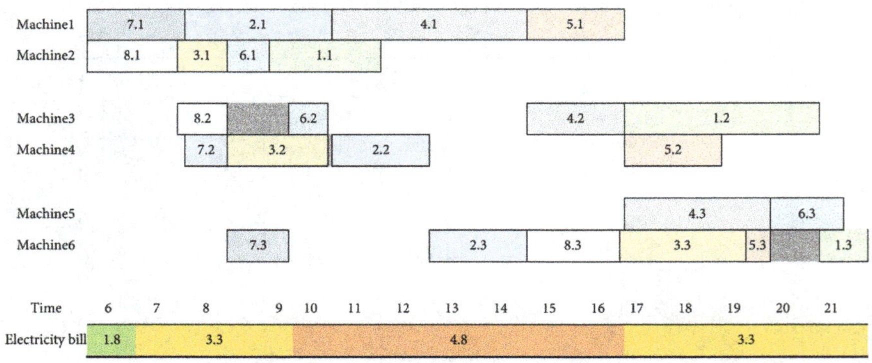

Figure 5. The Gantt chart of computational result using EGA.

Table 3. The electricity bill of time periods.

Factory working period		6:00~20:00
Time period		Summer electricity price (NT$)
Off-peak	6:00~7:00	1.8
Mid-peak	7:00~10:00	3.3
	17:00~20:00	3.3
On-peak	10:00~17:00	4.8

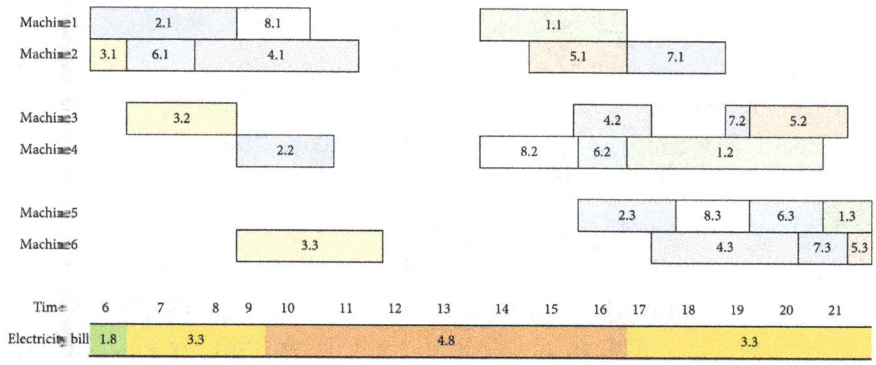

Figure 6. The Gantt chart of computational result using ACGA.

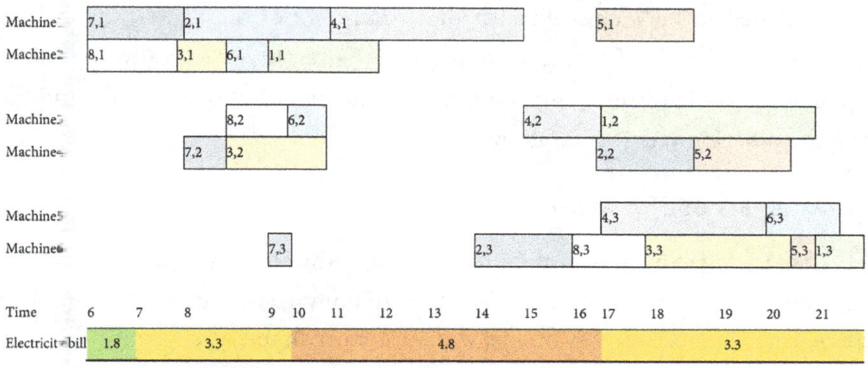

Figure 7. The Gantt chart of computational result using AEGA.

Table 4. Comparison of electricity bills.

	CGA	EGA	ACGA	AEGA
Electricity bills (NT$)	3277.8	2900.6	2929.7	2807.4
Comparison costs ratio	1	0.88	0.89	0.86

benchmark for comparison. According to **Table 4**, applying the CGA generates the highest electricity cost, because it neglects the costs of standby, shutdown, and different time periods. Furthermore, the EGA considers different conditions of machines and can lower the electricity cost; however, jobs are processed in the on-peak period. The proposed method using ESGA can efficiently avoid processing jobs in the on-peak period. Moreover, ESGA can lower the electricity cost by up to 11% and 14%, respectively, compared with the CGA. The results demonstrate higher energy saving with the AEGA than with the CGA, EGA, and ACGA.

4. Computational Experiments

All tests were conducted on a PC with an Intel® Core™ i5-4300U 1.9-GHz CPU with 8 GB of RAM. The operation system used was Windows® 8.1. The FFS scheduling problem is $FF_k \mid EP_t, State \mid EC \mid T_j = 0$. The parameters of the proposed method are the number of jobs, type of stage, number of machines, energy consumption of machines, processing time of jobs, and due dates. For the considered problems, the number of jobs was 30, 60, or 90; the type of stage was 3 or 5; and the number of machines in the two stages = 6, 9. **Table 2** and **Table 5** show the energy consumption of machines. Moreover, the processing time of jobs was limited within U [2] [9]; the unit time was 30 minutes; the factory working period was from 6 a.m. to 10 p.m. on each day; and the electricity costs of various time periods are shown in **Table 3**. The formula of the due date $(1-TF)\sum_{i=1}^{k}\overline{P}_i(1+(k+1)*0.25)$, among which the $\overline{P}_i$ represents the average processing time of machines, and TF is the tardy factor; Type I was 0.3 and Type II was 0.5. The working duration per day was assumed to be 16 hours. Therefore, when the formula of the due date provided a duration of 20 hours, the due date was estimated to be 2 days. Finally, the parameters of the genetic algorithms were set on the basis of a pretest to improve the results; specifically, the initial population was 30; the exchange rate was 0.9; the mutation rate was 0.1; and the stopping criterion was 500 iterations.

4.1. Analysis of Effectiveness

To assess the effectiveness of the proposed algorithm in solving the considered problems, 12 conditions were applied to randomly test 30 generated problem sets; specifically, the number of jobs was 30, 60, or 90; the type of stage was 3 or 5; and the number of machines in the two stages was 6 or 9. **Table 6** provided a comparison of the average solutions and average solving times of the CGA, EGA, ACGA, and AEGA for efficacy analysis under Type I of slacker due dates.

Table 7 provided a comparison of the average solutions and average solving times of the CGA, EGA, ACGA, and AEGA for efficacy analysis under Type II of non-slacker due dates.

Table 6 and **Table 7** illustrate that for different due dates, the cost of energy consumption is lower for the ACGA and AEGA than for the CGA. Type I has slacker due dates than those of Type II, indicating that the EGA and AEGA provide superior solutions. A data set of 30/6/5 has the same due date as that of Type I and Type II; therefore, the improvement is not remarkable. The CGA calculates the electricity cost for the minimal makespan; therefore, the CGA and ACGA might not provide superior solutions in the Type I condition. **Table 8** show comparison cost ratios for electricity costs obtained using all methods under Type I of slacker due dates.

Table 8 and **Table 9** show comparison cost ratios for electricity costs obtained using all methods under Type I of non-slacker due dates.

Table 5. The energy consumption during three time periods of 5-stage machines.

Conditions of machines	Stage 1	Stage 2	Stage 3	Stage 4	Stage 5
Stand-by (kW)	3	7	6	6	5
Operation (kW)	8	18	17	15	14
Turn-on (kW)	4	10	8	7	7

Table 6. The effectiveness of time window Type I.

	CGA		EGA		ACGA		AEGA	
j/m/k	Avg.	CPU time(s)	Avg.	CPU time(s)	Avg.	CPU time(s)	Avg.	CPU time(s)
30/6/3	12673	5.8	8558	7.1	12471	6.6	8314	7.5
30/6/5	22519	9.9	15957	12.1	20681	11.3	15438	12.6
30/9/3	13173	6.4	8721	7.6	12405	7.2	8616	8.2
30/9/5	23098	11.4	16516	13.8	20893	12.2	15706	14.4
60/6/3	26545	15.3	19420	17.6	24187	16.5	18242	18.2
60/6/5	44668	27.4	37244	32.4	42057	30.3	34822	33.2
60/9/3	25903	16	18763	19	24779	18.2	18412	19.7
60/9/5	44647	27	36474	31.7	42664	29.9	34236	32.1
90/6/3	39398	25.7	31909	29.4	37783	28.5	29634	30.8
90/6/5	65787	48.5	57154	53.9	62155	51.3	53657	55
90/9/3	40014	26.7	31456	30.5	37869	29.7	30431	31.1
90/9/5	66557	31.5	57340	37.4	64741	33.6	53933	38

Remarks: The unit cost is NT$.

Table 7. The effectiveness of time window Type II.

	CGA		EGA		ACGA		AEGA	
j/m/k	Avg.	CPU time(s)	Avg.	CPU time(s)	Avg.	CPU time(s)	Avg.	CPU time(s)
30/6/3	12700	5.7	8650	7.3	12523	6.7	8473	7.8
30/6/5	22243	10.2	16467	11.9	21347	10.9	15992	12.5
30/9/3	13041	6.4	8659	7.5	12597	7	8507	8.1
30/9/5	22773	11.3	17036	14.2	21965	12.7	16325	14.8
60/6/3	27427	15.1	20036	17.5	24911	16.8	18868	18.4
60/6/5	44884	27	37302	32.7	42566	29.6	36043	33.7
60/9/3	26008	15.6	19167	18.9	25348	17.8	18843	19.5
60/9/5	44988	27.5	36825	31.5	42497	30.2	35445	32.1
90/6/3	39286	25.3	33843	28.7	37859	27.6	31830	29.3
90/6/5	67006	47.1	57504	53	63318	51.9	55136	54.8
90/9/3	39850	27.2	33224	30.6	37368	29.4	32200	31.3
90/9/5	67166	32.4	57519	37	65489	34.2	54244	37.9

Remarks: The unit cost is NT$.

Table 8 and **Table 9** illustrate that the EGA, ACGA, and AEGA yielded lower electricity costs than did the CGA. Moreover, the AEGA provided solutions superior to those of the ACGA method in terms of effectiveness.

4.2. Analysis of Robustness

To assess the robustness of the proposed algorithms in solving the considered problems, 12 conditions were applied to test the same solution 30 times;

Table 8. The comparison costs ratio of electricity bills of time window Type I.

j/m/k	CGA	EGA	ACGA	AEGA
30/6/3	1	0.68	0.98	0.66
30/6/5	1	0.71	0.92	0.69
30/9/3	1	0.66	0.94	0.65
30/9/5	1	0.72	0.90	0.68
60/6/3	1	0.73	0.91	0.69
60/6/5	1	0.83	0.94	0.78
60/9/3	1	0.72	0.96	0.71
60/9/5	1	0.82	0.96	0.77
90/6/3	1	0.81	0.96	0.75
90/6/5	1	0.87	0.94	0.82
90/9/3	1	0.79	0.95	0.76
90/9/5	1	0.86	0.97	0.81
Avg.	1	0.77	0.94	0.73

Table 9. The comparison costs ratio of electricity bills of time window Type II.

j/m/k	CGA	EGA	ACGA	AEGA
30/6/3	1	0.68	0.99	0.67
30/6/5	1	0.74	0.96	0.72
30/9/3	1	0.66	0.97	0.65
30/9/5	1	0.75	0.96	0.72
60/6/3	1	0.73	0.91	0.69
60/6/5	1	0.83	0.95	0.80
60/9/3	1	0.74	0.97	0.72
60/9/5	1	0.82	0.94	0.79
90/6/3	1	0.86	0.96	0.81
90/6/5	1	0.86	0.94	0.82
90/9/3	1	0.83	0.94	0.81
90/9/5	1	0.86	0.98	0.81
Avg.	1	0.78	0.96	0.75

specifically, the number of jobs was 30, 60, or 90; the type of stage = 3, 5; and the number of machines in the two stages M_s = 6, 9. **Table 10** and **Table 11** provide a comparison of the average solutions, optimal solutions, and poorest solutions calculated using the CGA, EGA, ACGA, and AEGA in the analysis of robustness.

The test results in **Table 10** and **Table 11** for robustness show that the ACGA and AEGA are affected by original solutions during adjustment procedures,

Table 10. The robustness of time window Type I.

j/m/k	CGA Best / Worst	CGA Avg. [Std. dev.]	EGA Best / Worst	EGA Avg. [Std. dev.]	ACGA Best / Worst	ACGA Avg. [Std. dev.]	AEGA Best / Worst	AEGA Avg. [Std. dev.]
30/6/3	11330 12687	12202 [492.6]	8136 9105	8991 [356.6]	11083 12249	11693 [422.5]	7964 8832	8507 [307]
30/6/5	21853 23085	22469 [677.9]	16430 18034	17232 [537.5]	21090 22272	21836 [427.3]	15082 16512	16030 [420.3]
30/9/3	11336 13170	12253 [684.2]	8521 9118	8819 [353.6]	10534 12430	11707 [686.5]	8100 8864	8381 [497]
30/9/5	20937 23189	22169 [862.1]	16131 17282	16707 [457.5]	19170 22567	21395 [1304.3]	15434 16172	15805 [295.2]
60/6/3	24779 27195	25987 [984.5]	19118 20513	19822 [653.2]	23062 25432	24515 [936.9]	18267 19706	18945 [646.8]
60/6/5	42020 44735	43378 [999.2]	35442 36632	35805 [676.3]	39517 42368	41239 [1065.6]	34267 36301	35256 [758.8]
60/9/3	24565 27261	25913 [1149.6]	18209 19711	18920 [594.7]	24124 25826	24719 [646.8]	17415 18753	18288 [370.2]
60/9/5	43167 46538	44852 [1209.4]	35305 37244	36377 [881.6]	41856 43784	42481 [865.1]	33707 36761	35112 [845.2]
90/6/3	38541 40956	39749 [907.9]	32015 34513	33263 [955.7]	37124 38134	37712 [438]	27963 31085	29877 [1161.7]
90/6/5	65226 67744	66485 [929.0]	55357 57823	56175 [913.8]	61613 63426	62774 [720.1]	52056 54253	53456 [889.5]
90/9/3	39596 41817	40707 [839.8]	30663 31878	31270 [436.2]	35614 37759	37211 [941.5]	29920 31085	30545 [416.2]
90/9/5	65027 68482	66754 [1270.9]	55440 57461	56450 [682.9]	62526 66064	64235 [1878.7]	54006 55744	54620 [674.4]

Remarks: The unit cost is NT$.

causing the electricity cost to decrease partially. Thus, the robustness of the proposed method was fairly identified.

5. Computational Experiments

This study investigated the use of limited resources of parallel machines to promote practical production and environmental protection. Ideal practical production for better energy using effective scheduling prevents processes from excessive energy consumption and energy price fluctuations. Recent studies on sustainable manufacturing focused on energy saving to reduce the unit production cost and environmental impacts.

Most importantly, the application of the optimization of the scheduling can be respected for costs down and energy-saving simultaneously. Under an operational environment in which electricity costs differ depending on the time period, manufacturing activities in machine shops increase electricity consumption costs. In particular, machine conditions and job processing time periods are crucial factors in making energy-saving decisions during the manufacturing process.

Table 11. The robustness of time window Type II.

j/m/k	CGA Best / Worst	CGA Avg. [Std. dev.]	EGA Best / Worst	EGA Avg. [Std. dev.]	ACGA Best / Worst	ACGA Avg. [Std. dev.]	AEGA Best / Worst	AEGA Avg. [Std. dev.]
30/6/3	11882 13695	12654 [727.8]	8357 8995	8732 [285.6]	11719 13527	12501 [769.7]	8020 8990	8557 [346.5]
30/6/5	20613 23183	22795 [794.7]	16819 19539	18037 [939.6]	20512 22919	21555 [791.5]	15525 17692	16015 [983.6]
30/9/3	11685 13150	12432 [551.6]	8435 8886	8641 [257.9]	11414 13058	12287 652.9	8211 8831	8490 [239.5]
30/9/5	20870 22629	22116 [764.1]	16431 17863	17273 [483.5]	20720 22232	21755 [552.2]	15632 16812	16178 [444.5]
60/6/3	24996 27493	26199 [1040]	19632 20944	20292 [409.6]	23821 25713	24872 [741.9]	18579 20105	19216 [563.4]
60/6/5	42195 45796	43873 [1267.5]	35972 37976	36790 [500.5]	40297 44283	42243 [1425.9]	34933 37009	35929 [657.7]
60/9/3	25168 27220	26187 [747.1]	19257 20980	20118 [634.1]	24323 26260	25308 [760.3]	17821 19056	18603 [743.8]
60/9/5	42133 46152	44225 [1551.5]	35563 38106	37033 [1311.8]	40184 44258	42705 [1595.9]	34534 36907	35509 [1163.7]
90/6/3	38663 41731	39708 [1228.7]	31907 34748	33670 [1059.8]	37485 38319	37807 [341.1]	30927 34023	32197 [1393.3]
90/6/5	65131 67830	66630 [1007.9]	56002 58882	57442 [1111.8]	62199 65876	64249 [1429.4]	53741 56412	54987 [1180]
90/9/3	38270 40291	39270 [720.2]	32262 34060	33093 [664.9]	36261 38974	37340 [918.2]	30489 32286	31587 [649.5]
90/9/5	66374 68401	67386 [737.8]	55781 58106	57005 [896.2]	63054 66193	64489 [1320.7]	54669 56518	55593 [779]

Remarks: The unit cost is NT$.

According to the analysis, the CGA, EGA, ACGA, and AEGA can efficiently solve problems. The electricity cost ratio for the CGA, EGA, ACGA, and AEGA is 1:0.78:0.95:0.74, demonstrating that the AEGA can efficiently solve the problem. The robustness test results show that the ACGA and AEGA of ESGA are affected by original solutions during adjustment procedures and partially reduce electricity costs. Thus, the robustness of the proposed method was appropriately identified. As a whole, the proposed method can lower electricity bills to fit green energy nowadays.

Finally, enterprises can adopt the proposed method for enhancing practical production in the flexible job shop scheduling environment and generate profits by fully exploiting the advantages of the method in the future. In addition, the proposed method can mitigate the environmental impact of manufacturing processes and protect environment at the same time.

References

[1] Goldratt, E.M. and Cox, J. (1984) The Goal: Excellence in Manufacturing. North River Press, Great Barrington, MA.

[2] Tongur, V. and Ülker, E. (2014) Migrating Birds Optimization for Flow Shop Sequencing Problem. *Journal of Computer and Communications*, **2**, 142-147. https://doi.org/10.4236/jcc.2014.24019

[3] Zabihzadeh, S.S. and Rezaeian, J. (2016) Two Meta-Heuristic Algorithms for Flexible Flow Shop Scheduling Problem with Robotic Transportation and Release Time. *Applied Soft Computing*, **40**, 319-330. https://doi.org/10.1016/j.asoc.2015.11.008

[4] Linn, R. and Zhang, W. (1999) Hybrid Flow Shop Scheduling: A Survey. *Computers and Industrial Engineering*, **37**, 57-61. https://doi.org/10.1016/S0360-8352(99)00023-6

[5] Ribas, I., Leisten, R. and Framiñan, J.M. (2010) Review and Classification of Hybrid Flow Shop Scheduling Problems from a Production System and a Solutions Procedure Perspective. *Computers & Operations Research*, **37**, 1439-1454. https://doi.org/10.1016/j.cor.2009.11.001

[6] Dai, M., Tang, D., Giret, A., Salido, M.A. and Li, W.D. (2013) Energy-Efficient Scheduling for a Flexible Flow Shop Using an Improved Genetic-Simulated Annealing Algorithm. *Robotics and Computer-Integrated Manufacturing*, **29**, 418-429. https://doi.org/10.1016/j.rcim.2013.04.001

[7] Huang, R.H., Yu, S.C. and Kuo, C.W. (2014) Reentrant Two-Stage Multiprocessor Flow Shop Scheduling with Due Windows. *The International Journal of Advanced Manufacturing Technology*, **71**, 1263-1276. https://doi.org/10.1007/s00170-013-5534-4

[8] Mukund Nilakantan, J., Huang, G.Q. and Ponnambalam, S.G. (2015) An Investigation on Minimizing Cycle Time and Total Energy Consumption in Robotic Assembly Line Systems. *Journal of Cleaner Production*, **90**, 311-325. https://doi.org/10.1016/j.jclepro.2014.11.041

[9] Zhang, H., Zhao, F., Fang, K. and Sutherland, J.W. (2014) Energy-Conscious Flow Shop Scheduling under Time-of-Use Electricity Tariffs. *CIRP Annals-Manufacturing Technology*, **63**, 37-40. https://doi.org/10.1016/j.cirp.2014.03.011

[10] Setlhaolo, D., Xia, X. and Zhang, J. (2014) Optimal Scheduling of Household Appliances for Demand Response. *Electric Power Systems Research*, **116**, 24-28. https://doi.org/10.1016/j.epsr.2014.04.012

[11] Jin, W., Hu, Z. and Chan, C. (2017) An Innovative Genetic Algorithms-Based Inexact Non-Linear Programming Problem Solving Method. *Journal of Environmental Protection*, **8**, 231-249. https://doi.org/10.4236/jep.2017.83018

[12] Mouzon, G., Yildirim, M.B. and Twomey, J. (2007) Operational Methods for Minimization of Energy Consumption of Manufacturing Equipment. *International Journal of Production Research*, **45**, 4247-4271. https://doi.org/10.1080/00207540701450013

[13] Fang, K., Uhan, N., Zhao, F. and Sutherland, J.W. (2011) A New Approach to Scheduling in Manufacturing for Power Consumption and Carbon Footprint Reduction. *Journal of Manufacturing Systems*, **30**, 234-240. https://doi.org/10.1016/j.jmsy.2011.08.004

[14] Tibi, N.A. and Arman, H. (2007) A Linear Programming Model to Optimize the Decision-Making to Managing Cogeneration System. *Clean Technologies and Environmental Policy*, **9**, 235-240. https://doi.org/10.1007/s10098-006-0075-2

[15] He, Y., Li, Y.F., Wu, T. and Sutherland, J.W. (2015) An Energy-Responsive Optimization Method for Machine Tool Selection and Operation Sequence in Flexible Machining Job Shops. *Journal of Cleaner Production*, **87**, 245-254. https://doi.org/10.1016/j.jclepro.2014.10.006

[16] Bruzzone, A.A.G., Anghinolfi, D., Paolucci, M. and Tonelli, F. (2012) Energy-Aware Scheduling for Improving Manufacturing Process Sustainability: A Mathematical Model for Flexible Flow Shops. *CIRP Annals-Manufacturing Technology*, **61**, 459-462. https://doi.org/10.1016/j.cirp.2012.03.084

[17] Shao, W., Pi, D. and Shao, Z. (2017) Memetic Algorithm with Node and Edge Histogram for No-Idle Flow Shop Scheduling Problem to Minimize the Makespan Criterion. *Applied Soft Computing*, **54**, 164-182. https://doi.org/10.1016/j.asoc.2017.01.017

[18] Lian, Z.G. (2010) A Local and Global Search Combine Particle Swarm Optimization Algorithm for Job-Shop Scheduling to Minimize Makespan. *Discrete Dynamics in Nature and Society*, **2010**, Article ID 838596, 11 p.

[19] Zhang, C.S., Ning, J. and Ouyang, D. (2010) A Hybrid Alternate Two Phases Particle Swarm Optimization Algorithm for Flow Shop Scheduling Problem. *Computers and Industrial Engineering*, **58**, 1-11. https://doi.org/10.1016/j.cie.2009.01.016

[20] Liu, H.C., Gao, L. and Pan, Q. (2011) A Hybrid Particle Swarm Optimization with Estimation of Distribution Algorithm for Solving Permutation Flowshop Scheduling Problem. *Expert Systems with Applications*, **38**, 4348-4360. https://doi.org/10.1016/j.eswa.2010.09.104

[21] Zhao, F., Tang, J., Wang, J. and Jonrinaldi, J. (2014) An Improved Particle Swarm Optimisation with a Linearly Decreasing Disturbance Term for Flow Shop Scheduling with Limited Buffers. *International Journal of Computer Integrated Manufacturing*, **27**, 488-499. https://doi.org/10.1080/0951192X.2013.814165

[22] Li, T., Castro, P.M. and Lv, Z. (2017) Life Cycle Assessment and Optimization of an Iron Making System with a Combined Cycle Power Plant: A Case Study from China. *Clean Technologies and Environmental Policy*, **19**, 1133-1145. https://doi.org/10.1007/s10098-016-1306-9

[23] Ma, B. and Yu, Y. (2017) Industrial Structure, Energy-Saving Regulations and Energy Intensity: Evidence from Chinese Cities. *Journal of Cleaner Production*, **141**, 1539-1547. https://doi.org/10.1016/j.jclepro.2016.09.221

[24] Peyramalea, V. and Wetzelb, C. (2017) Analyzing the Energy-Saving Potential of Buildings for Sustainable Refurbishment. *Procedia Environmental Sciences*, **38**, 162-168. https://doi.org/10.1016/j.proenv.2017.03.098

[25] Moon, J.Y., Shin, K. and Park, J. (2013) Optimization of Production Scheduling with Time-Dependent and Machine-Dependent Electricity Cost for Industrial Energy Efficiency. *The International Journal of Advanced Manufacturing Technology*, **68**, 523-535. https://doi.org/10.1007/s00170-013-4749-8

[26] Shrouf, F., Ordieres-Mere, J., Garcia-Sanchez, A. and Ortega-Mier, M. (2014) Optimizing the Production Scheduling of a Single Machine to Minimize Total Energy Consumption Costs. *Journal of Cleaner Production*, **67**, 197-207. https://doi.org/10.1016/j.jclepro.2013.12.024

[27] Pinedo, M.L. (2008) Scheduling: Theory, Algorithm and Systems. 3rd Edition, Springer-Verlag, New York.

2

Measurements and Visualization of the Fluid Field of the Plume from an Animal Housing Ventilation Fan

Manqing Ying[1], Lingjuan Wang-Li[1*], Larry F. Stikeleather[1], Jack Edwards[2]

[1]Department of Biological and Agricultural Engineering, North Carolina State University, Raleigh, USA
[2]Department of Mechanical and Aerospace Engineering, North Carolina State University, Raleigh, USA
Email: *lwang5@ncsu.edu

Abstract

Various dispersion models have been developed to simulate the fate and transport of air emissions from animal housing systems to meet the increasing need for knowledge in this area. However, the accuracy of the models may be challenged due to the unknown plume rise and plume shape. This paper reports a combination of theoretical and field study of the plum rise and shape of air flow from a ventilation fan commonly used in mechanically ventilated animal houses. The theoretical modeling of the plume shape was conducted using a commercial Computational Fluid Dynamics (CFD) package named FloEFD; the field measurements of the plume field was conducted using five 3D ultrasonic anemometers to simultaneously measure the air flow in the plume at various locations (four heights and five downwind distances). The TECPLOT package was used to visualize the plume flow field based upon anemometer measurements. While the plume shapes were found to be left-shifted by the CFD model and TECPLOT visualization, the magnitudes of the 3D wind velocities from field measurement were found to be significantly larger than those from CFD model. The plume field measurements indicated that the plume of a 0.6 m (24-inch) ventilation fan had a depth about 9 m, a width about ±6 m, and a rise (lifting) beyond the highest measurement point, 4.88 m (16 ft).

Keywords

Computational Fluid Dynamics (CFD), Plume Fluid Field Measurement, Plume Rise, Animal Housing

1. Introduction

Animal products are more and more in need to feed the rapidly growing population. It has been reported [1] that there were 450,000 animal feeding operations (AFOs) in the United States. Along with the significant economic and social benefits produced by the AFO industry, however, serious environmental concerns have been raised due to massive production of animal waste and air emissions from the operations. In assessment of impact of AFO air emissions on local and regional air quality, knowledge of the fate and transport of those emissions is required, and yet has not been well studied and understood.

While estimations of air emissions from AFO facilities have been studies intensively [2] [3] [4], investigation of the fate and transport of the emissions is a relatively newer research topic in recent years [5] [6] [7] [8]. In the literature, estimations of downwind pollutant concentrations in responses to AFO emissions have been commonly conducted through Gaussian dispersion modeling approach [8] [9] [10]. The Gaussian dispersion model was originally developed based upon observations of industrial "stack" emissions with vertical momentum and thermal buoyancy [11]. The accuracy of the Gaussian dispersion model may be challenged when it is applied for assessing dispersion of air emissions from animal housing system where there is lack of vertical momentum and sufficient "stack" height. This challenge mainly comes from unknown of plume rise and plume shape of the horizontal emissions from the housing ventilation fans.

The rise of the centerline of a dispersing plume is called the plume rise. Currently, the Briggs' formula and the Holland's formula [11] are most commonly used models to estimate the plume rises. Despite the fact that the estimation of these two plume rise models do not agree with each other, both the Briggs' formula [5] and the Holland's formula compute the plume rise by two parts: the rise caused by thermal lifting and the rise caused by vertical momentum.

Applying both Briggs and Holland formulas into the case of animal housing systems, a small plume rise would be expected due to the lack of vertical momentum and thermal buoyancy under warm weather condition. However, field observation showed that the plume rise of exhaust air from animal buildings could actually reach as high as 10 m (**Figure 1**). Moreover, unlike the plume rise

Figure 1. Plume of air emissions (mixture of PM + gas + moisture) from two tunnel ventilated poultry houses (Note: the background trees were over 14 m high) (image by Wang-Li).

models for industrial stacks used in the existing formulas, for AFO housing system, the emission "stacks" are much lower with horizontal initial air flow instead of vertical flow as compared to the industrial stacks. Thus, the accuracy of the existing formulas is challenged when applying to the case in animal housing system.

In an effort to assess an emission plume, Holmes *et al.* [12] reported that the flow patterns of gas and PM emissions agreed well in open environments unless there was severe turbulence or the emission sources were too complicated. Understanding of fluid filed of the exhaust plume from animal housing ventilation fans may lead to quantification of rises of the emission plumes. In theory, fluid dynamic equations, which are mostly in the forms of partial differential equations, may be used to express the fundamental principles of any fluid field and to predict fluid field behavior under any given boundary conditions [13]. To find a numerical description of the fluid field that meets the fundamental principles and the given boundary conditions, as much as millions of iterations could happen. To facilitate such high computational demand, the computational fluid dynamics (CFD) has been commonly used in fluid field calculation and modeling. In application of the CFD modeling, fluid field measurements are usually required to validate and calibrate the model outputs. While there are many publications about both CFD modeling and field measurements of industrial fan flows [14] [15] [16], there is seldom comparison between them, especially for closer downwind fluid field studies. This is probably because, as Quinn *et al.* [17] concluded, the performance of the modeling systems depends highly on the fluctuation of the wind conditions.

In addition to CFD method for fluid field modeling, TECPLOT [18] is another fluid dynamic package to visualize and model the fluid field behavior when the fluid field measurements are available. It has been applied in visualization of various research fields such as heat transfer process [19], flow field measurement [20] and even chemical reactions [21].

This study compared the flow filed measurements with the CFD model outputs of the fluid field of the exhaust plume from a ventilation fan commonly used in AFO housing systems. The specific objectives of the research was to 1) measure the fluid field of the exhaust plume from an animal housing ventilation fan; 2) visualize and quantify the shape and rise of the plume through CFD modeling and TECPLOT visualization.

2. Methodology

In this study, two approaches were taken to model and visualize rise of the plume emitted from a ventilation fan testing setup. The first approach measured fluid field of the ventilation plume under the controlled setting. The fluid field measurements were then applied to TECPLOT to visualize the exhaust plume shapes defined by plume height, depth and width. In the second approach, a commercially available CFD package (FloEFD 11 for Creo, Mentor Graphics) was used to simulate the fluid field of the plume emitted from the ventilation fan testing setup. Through the plotting of the model resulted by CFD approach, plume rise and shape was then observed and quantified.

2.1. Ventilation Plume Testing Setup: The Mini-Tunnel

To be consistent with field observation, a wooden mini-tunnel with an axial ventilation fan was constructed to simulate ventilation flow that would be typically observed in a poultry housing system. This testing mini-tunnel has a dimension in 1.22 m × 0.91 m × 2.44 m (4 ft × 3 ft × 8 ft) with an axial ventilation fan (0.6 m, 24" in diameter, AT24ZCP, Aerotech) installed on the one end. To ensure similar air flow pattern in the testing tunnel as it is in a typical poultry house, a collimating screen was installed at the entrance end of the tunnel. This collimating screen not only stabilized the airflow in the tunnel with short traveling distance, but also generated a pressure drop around 12.50 Pa (0.05 in-H_2O) (**Figure 2**). The wooden mini-tunnel was placed in an open area on the Lake Wheeler Field Laboratory of North Carolina State University for the plume testing.

2.2. Ventilation Plume Measurements

2.2.1. Instrumentation

The plume field velocity measurements were conducted using five 3D ultrasonic anemometer assemblies with adjustable heights (**Figure 3**). Each assembly consisted of a 3D anemometer (Model 81,000, R. M. Young Company, Traverse City, Michigan), a battery and a 4-channel data logger (Onset U12-006, Onset Computer Corporation, Cape Cod, MA) to record measurements of 3D velocities and temperature. The data logger can store 14 hours of data with a 10 second recording interval.

Before the field testing, some default settings of the anemometers were changed to fit the experimental design. These changes include:

- "serial output form"—changed to "u, v, w", "Ts (sonic temperature)" and "internal voltage";
- The velocities were denoted by u_0, v_0 and w_0, among which $+u_0$ values = wind from the east; $+v_0$ values = wind from the north; $+w_0$ = wind from below (updraft).

Figure 2. The wooden mini-tunnel with a 0.6 m (24-in) axial ventilation fan on the one end and the collimating screen on the other end.

Figure 3. The 3D ultrasonic anemometer assemblies (left: anemometer on the post with data logger and battery power supply) and illustration of the adjustable heights of the anemometer assembly on the post (right).

- "voltage output format"—"scaling" was changed to 15 m/s.

In addition to the tunnel plume field measurements, the local meteorological data were monitored to count for background wind effect. This meteorological data collection was conducted using a 10 meter weather tower installed about 30 meters away from the mini-tunnel testing field. Wind speed, wind direction and solar radiation, air temperature, RH were monitored by the sensors on the tower (Onset Computer Corporation, Cape Cod, MA) including a wind speed/direction smart sensor (S-WCA-M003), a solar radiation shield (RS3), a temperature/RH smart sensor (S-THB-M00x), a multi-channel logger (HOBO U30, Onset Computer Corporation, Cape Cod, MA) to record data every minute and a U-shuttle (U-DT-2) for reading data from the logger and transfer to a host computer.

2.2.2. Plume Fluid Field Measurements

Before the measurements of the plume fluid field, some preliminary smoke tests were done to visualize the width and depth of the plume for placement of the anemometers. The preliminary smoke tests showed that the plume at full rpm capacity went as far as 15 m before rising up and could reach as high as 4.57 m (15 ft) after rising up and bending over. Based upon the observations of the smoke tests, the plume fluid field was divided into 5 rings (3 m, 6 m, 9 m, 12 m, 15 m) away from the fan and 4 heights (0.30 m = 1 ft, 1.22 m = 4 ft, 3.05 m = 10 ft, 4.88 m = 16 ft). The top view of field testing is displayed in **Figure 4**. As the flow generated by the fan was constant at a given fan flow rate (rpm), the exhaust flow field was considered unchanged when the impact of background wind was subtracted. Thus, the plume field air velocity measurements were conducted one ring one height at a time.

For first set of tests, measurements at five locations were simultaneously taken at one height in one ring in each test with data collection for 2 hours at 10 second interval. Total of 12 tests were conducted for 4 heights (0.30 m = 1 ft, 1.22 m = 4 ft, 3.05 m = 10 ft, 4.88 m = 16 ft) and 3 rings (3 m, 9 m, 15 m). **Figure 5** illustrates the setup for one test at the height of 3.05 m (10 ft).

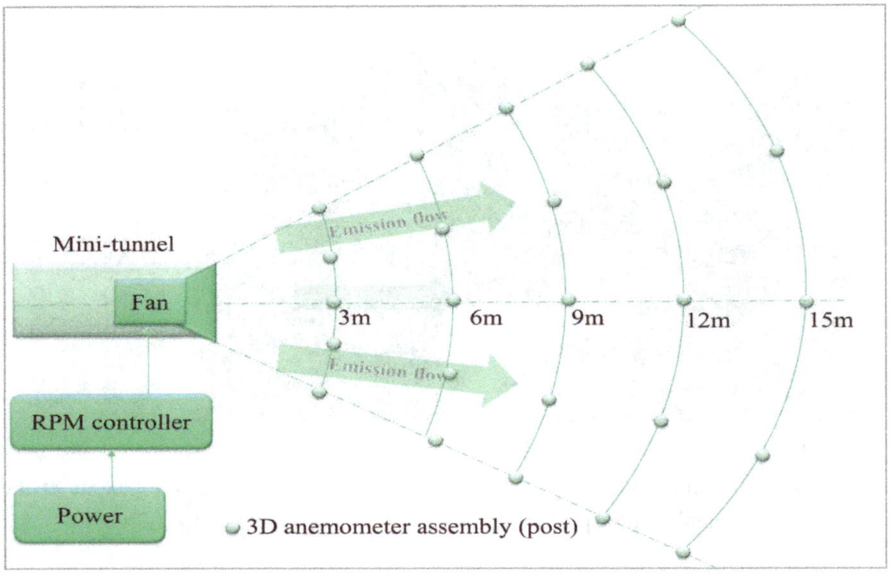

Figure 4. Top view of the 3D plume flow measurement layout.

Figure 5. The plume flow field measurement on the 3 m ring and at 3.05 m (10 ft) height.

For the second set of tests, the experimental design was the same as the first set of tests. Thus, another 12 tests were conducted for 4 heights (0.30 m = 1 ft, 1.22 m = 4 ft, 3.05 m = 10 ft, 4.88 m = 16 ft) and 3 rings (3 m, 9 m, 15 m).

In these two sets of tests, background wind data measured by the weather station on the 10 m tower were used to count for background wind effect on the plume filed velocity measurements.

In the third set of tests, to be more precisely factor out the background wind effect on the plume flow measurements, the background wind measurements in the testing field were also conducted. For each test, plume field 3D air velocities were taken for 10 minutes when the fan was on, then, the background 3D wind velocities in the same field were taken for another 10 minutes when the fan was turned off. In this set, total of 20 tests were conducted for 4 heights (0.30 m = 1 ft, 1.22 m = 4 ft, 3.05 m = 10 ft, 4.88 m = 16 ft) and 5 rings (3 m, 6 m, 9 m, 12 m, 15 m).

2.3. Data Processing

2.3.1. Flow Vector Transformation

To better describe plume shape (*i.e.* depth, width and rise) in line with the testing setup, a coordinate system was built based on the testing tunnel/fan position to transfer 3D velocity measurements from the North-South and West-East coordinates to the new coordinate system. As shown in **Figure 6**, in the new coordinate system, the +x was aligned with the fan pointing towards the flow direction; the +y was perpendicular to the x axial toward the left side of the fan; the +z was perpendicular to the x-y plane pointing upward from the ground. Based on the historical weather records, the prevailing wind at the field was in south-west direction, thus the fan was aligned 57 degrees clockwise from the north direction (**Figure 6**).

The 3D velocities measured by the anemometers and the background wind velocities measured by the weather tower were transferred to the coordinate system by:

$$u = -u_0 \cdot \sin 57^{\circ} - v_0 \cdot \cos 57^{\circ} \tag{1}$$

$$v = -v_0 \cdot \sin 57^{\circ} + u_0 \cdot \cos 57^{\circ} \tag{2}$$

The 3D velocity measurements were then averaged to every 10 minutes. For the third set of tests, by subtracting 10 min background 3D background wind velocity measurements (fan off), the measurement of the plume flow field as impacted only by the fan was resulted. For the first two sets of test, the background 2D (east-west & north-south) wind velocities measured by the weather tower were used for subtraction to obtain the plume field measurements as impacted only by the ventilation fan.

2.3.2. Weather Data Processing

For the first two sets of tests, since no in field background wind was measured,

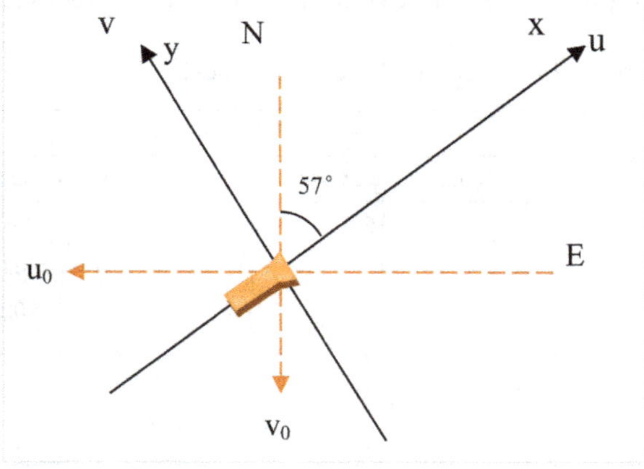

Figure 6. The coordinate system for plume flow field velocity computation.

the wind speed/direction data from the weather station were used. As the weather data were recorded every minute, it was first averaged to every 10 minutes. Thus, u_0's and v_0's (**Figure 6**) were averaged separately and were transformed by Equations ((1) and (2)) to u's and v's (**Figure 6**) at the measurement height of 10 meter. The measured wind directions were assumed to be within the horizontal plane, which suggested that there was no vertical impaction from the weather. Appling the power law [11], the wind speeds in two directions (u & v) were then converted to the speeds at the heights of the 3D anemometers at the time for taking the plume field measurements. The power law takes the following form:

$$\frac{u_1}{u_2} = \left(\frac{z_1}{z_2}\right)^p \qquad (3)$$

where,

z_1, z_2 = elevations 1 and 2, in this study, elevations of the 3D anemometers and the tower, 10 m.

u_1, u_2 = wind speeds at z_1 and z_2

p = exponent

The power p in Equation (3) varies with atmospheric stability class and surface roughness. It may be determined from the **Table 1** below:

In this study, the smooth surface was chosen for calculation determination. Turner's [22] method has been used to define atmospheric stability classes. This method is the revised version of the most commonly used Pasquill-Gifford method [23] [24]. The classification of the stability classes requires information about the solar altitude, wind speed and the cloud cover. Detailed procedure for determining the stability classes are reported in Cooper and Alley [11]. After transforming the weather data, the converted wind velocities were used as background wind for plume filed final flow calculation to identify plume profile as impacted solely by the ventilation fan.

2.4. Plume Fluid Field 3D Velocity Plotting: The Techplot Visualization

After subtracting the background wind impact for all the tests, the resulted 3D

Table 1. Exponents for Wind Profile (Power Law) Model.

Stability Class	Exponent (p)	
	Rough Surface (urban)	Smooth Surface (rural)
A	0.15	0.07
B	0.15	0.07
C	0.20	0.10
D	0.25	0.15
E	0.30	0.35
F	0.30	0.35

Adapted from U.S. Environmental Protection Agency, 1995.

velocities were considered to be the plume caused simply by the fan flow. Thus, the flow field presented by the 3D velocities should be symmetric.

Once the 10 minute averages of 3D velocity obtained for each set of tests, a plot of velocity vectors and contour of the measured plume field was generated by the software-TECPLOT 360, 2011 (Tecplot, Inc. Bellevue, Washington). Procedure for plotting in TECPLOT include the following:

1) sorting the data into 6 columns, denoting "x", "y", "z", "u", "v", "w", respectively;

2) exporting the data into a text file with the first line "variables = x, y, z, u, v, w" and the second line "zone i = 3 for test set one and two, 5 for test set three, j = 5, k = 4";

3) after the first two lines would be the 6 columns of variables;

4) importing the text file into the TECPLOT;

5) customizing the plot and generating contour map of the fluid field.

2.5. Plume Fluid Field CFD Modeling

The commercial CFD software, FloEFD was used to simulate the plume flow field of the axial fan by solving the governing equations. Other than the Navier-Stokes equations for mass, momentum, and energy conservation laws, specific equations describing the fluid could be customized by users based on the study case.

The code was developed to solve both laminar and turbulent flows. FloEFD used the *k*-ε model to express the transport equations for the turbulent kinetic energy and its dissipation rate. In this study, as a common case was expected to be simulated, the settings of the turbulence parameters were kept as default (C_μ = 0.09, $C_{\varepsilon 1}$ = 1.44, $C_{\varepsilon 2}$ = 1.92, σ_k = 1.0, and σ_ε = 1.3).

2.5.1. Definition of the Boundary Conditions

As no plume flow was defined in the model, the boundaries of the field were the ground, the fan and the fan discharge cone. The default type of the boundaries was adiabatic. Since there is no temperature difference between the airflow and the ambient air, the default setting in temperature was used for all the boundaries. As the ground in the field was covered with grass, the roughness of the boundaries was set to be 0.03 m as commonly used [25].

The fan information was imported from the engineering database inside FloEFD. The engineering database allowed the users to customize the fan. The input data are listed in **Table 2**:

The dimensions of the fan cone are 0.66 m in depth, 0.62 m in inter diameter,

Table 2. Settings of the ventilation fan in the FloEFD model.

Fan outer diameter	Hub diameter	Volume flow	Rotor speed	Rotation	Fan curve*
0.62 m	0.15 m	2.87 m^3/s	112.57 rad/s	clockwise	Q vs. □P

*The fan curve represents the static pressure versus the volume airflow and was tested at the BioEnvironmental and Structural Systems (BESS) Lab at University of Illinois at Urbana-Champaign (#03033).

Table 3. Settings of the computational domain in the FloEFD model.

Min x	Max x	Min y	Max y	Min z	Max z
−1 m	19 m	−20 m	20 m	0 m	30 m

and 0.79 m in outer diameter. As the cone was made of plastic, the roughness of the surface was set to be 0.0015 m.

2.5.2. Computational Mesh

A computational domain was first developed to define the region of calculation. In this study, the origin was set on the centerline of the fan system right at the edge of the discharge cone. The orientation of the coordinate system is the same as the one in the field measurement. The edges of the computational domain are listed in **Table 3**:

The refinement of the computational mesh was conducted automatically during the simulation process. The principle was to evenly split the current mesh to 8 child cells when interface was observed until the mesh size reached the threshold defined by the users. In this study, the mesh refine degree was selected to be 7 (5 to 10, greater number means smaller threshold) as 7 was enough to show the results while the finer mesh settings would take much longer time.

2.5.3. Simulation and Comparison

The end of simulation was set to be after 100 iterations as the mean and maximum velocities did not vary a lot after that. The model outputs were then obtained.

A plane parallel to the ground was set to view the results of the model at various heights. Then probe tool was used to capture the 3D velocities at the points where the anemometers in the field were set. Paired two-sample t-tests were done to analyze the difference between the model and one set of field data in winter. Also, the velocities observed by the probes in CFD were recorded in the text and excel file and plotted in TECPLOT.

3. Results and Discussion

3.1. Field Measurement

Due to some sensor and logger connection problem, for the first set of tests, only 8 groups of measurements were valid and named as group 1-01 through 1-08. For the second set of tests, 12 groups of results were valid and named group 2-01 through 2-12. Each group had 60 data points of 10-minute average 3D velocities. For the third set of tests, 1 scenario was generated with 100 points of 10-minute average 3D velocities and named as group 3.

Velocity Measurement Comparison within the First Test Set and Second Test Set

The comparisons of velocities among the groups within the first and second test sets were conducted by the statistical software R. Tukey HSD test was applied to

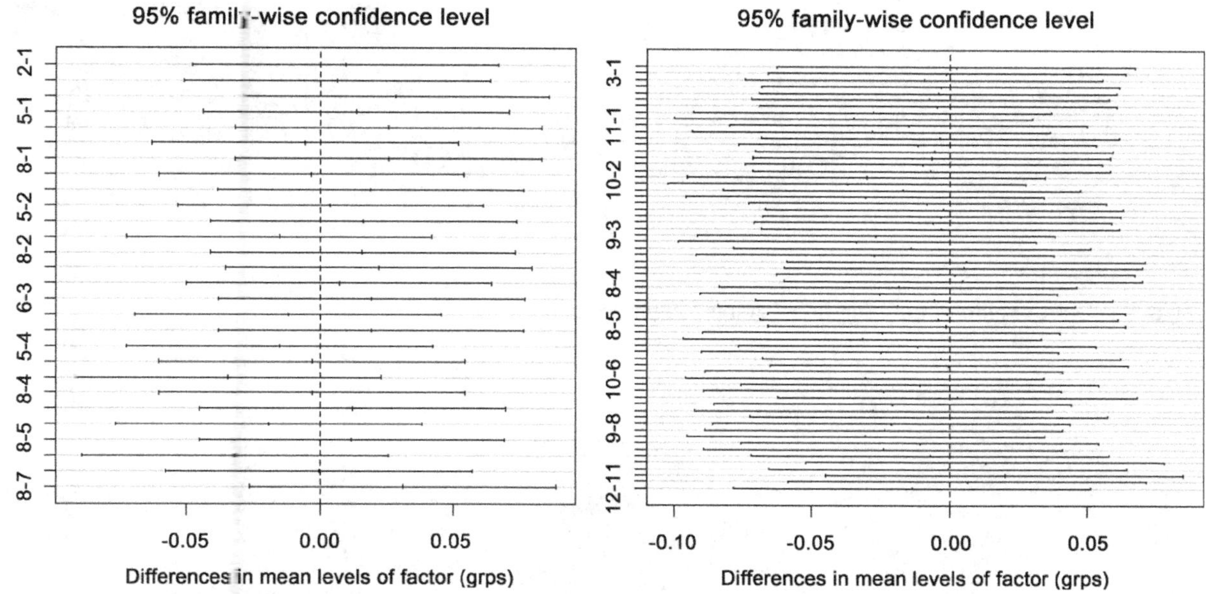

Figure 7. Comparisons of velocity in w direction (m/s) within first set of tests (left) and second set of tests (right).

compare the means of the velocities in three directions respectively. **Figure 7** shows an example of comparisons of the velocity in W direction (vertically up) for the first and second test sets.

The Tukey HSD tests revealed that there were no significant differences in u, v, w velocities among groups in both test sets.. The flow velocities in the vertical direction, w, was chosen as a representative because it was related to the plume lifting, thus to the rise of the plume, which is the target parameter this research is investigating.

The same TukeyHSD tests were also conducted one u, v and w between test set one and test set two. The results showed no significant differences. This indicated that the adjustment of the flow field measurements by subtracting the background wind in horizontal plane did not significantly change the consistence of the flow field measurements.

3.2. Plume Visualization by Techplot

TECPLOT was used to generate the 3D contour plot for each group of data. As the vertical lifting was the main focus, the velocity in vertical direction, w's were plotted for visualization of the plume field for all the testing groups. Since no significant difference was found among tests within sets one and two, randomly selected plots are shown in **Figures 8-11** for illustration. These figures show the displays of the 3D domain from different angles. By the "slices" function in TECPLOT, planes could be inserted to display the contour plot at certain planes.

The plotting results showed that the plume was shifted to the left in general. This was different from the initial expectation of a symmetric plume when there was no background wind interference. The measurement based plotting results for all three sets of tests show a left-shifted plume. Fact is that all the results from the three sets of test were point towards left.

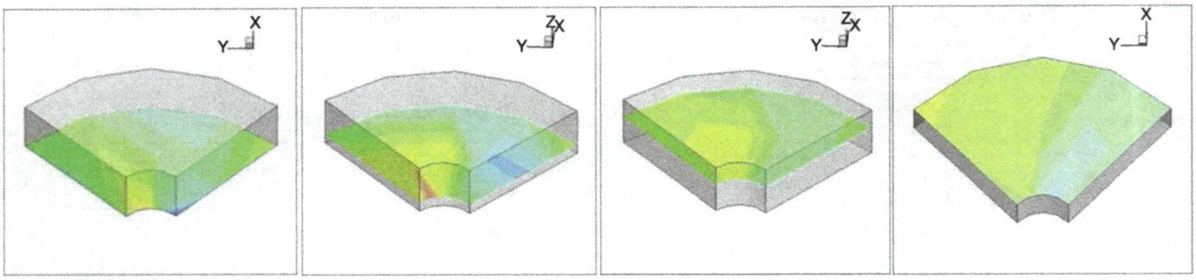

Figure 8. Contour plots of the velocity "w" at four heights for group 2-05.

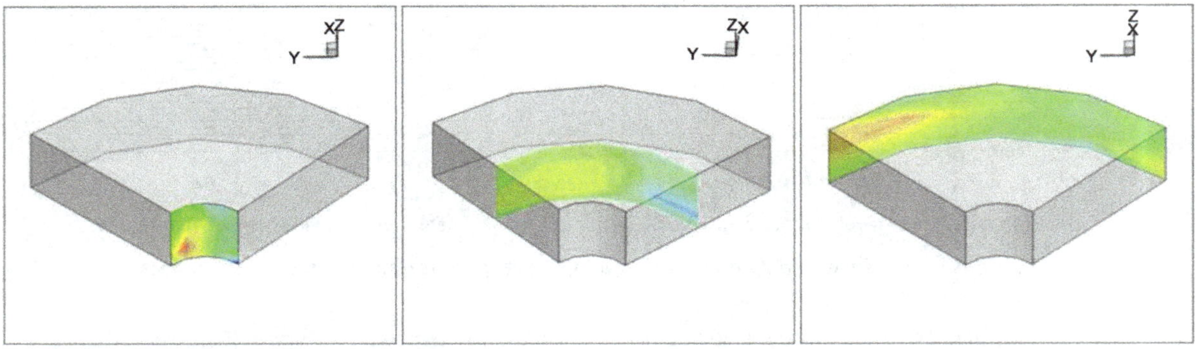

Figure 9. Contour plots of the velocity "w" at three rings (3 m, 6 m, 9 m) for group 2-05.

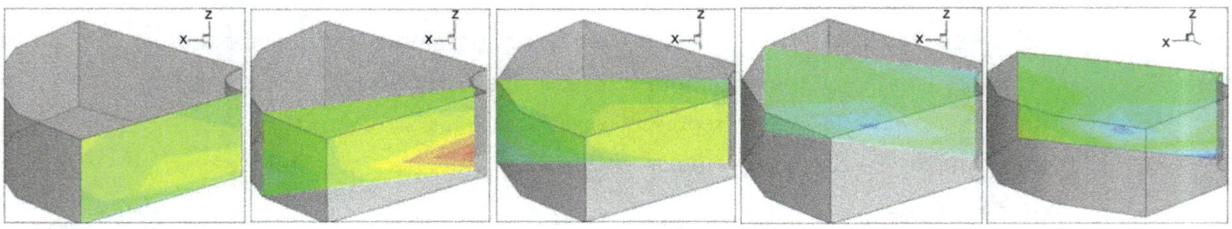

Figure 10. Contour plots of the velocity "w" at five angles (side views) for group 2-05.

While the first two sets showed reasonable flow patterns, the contour plots of test set three did not show a shape of the plume. Meanwhile, paired two-sample t-test between test set three and test set two indicated significant differences between the two. This might be due to the variation of the background wind since the background wind and fan flow velocities were not measured simultaneously. Since only one group of test was done for the test set three, further experiments should be conducted to justify this method.

3.3. Plume Modeling and Visualization by CFD

The CFD modeling was conducted for the given boundary condition specified in **Table 2** and **Table 3**. Upon obtaining the CFP outputs, paired two-sample t-test was used to test the difference between the CFD simulation and the field measurement. The results showed significant differences for comparing the model outputs in three velocities (u, v, w) with all groups' measurements of the velocities. While the CFD model produced velocities in much smaller magnitude, it showed a similar flow direction, shifting to the left. The top-view of the plume at 0.3 m above the ground and the computational meshes modeled by FloEFD is presented in **Figure 11**.

The contour of the CFD model also provided similar velocity profile as illustrated by the TECPLOT, but with a smaller magnitude. As an example, **Figure 12** shows left side view of the CFD contour plot.

3.4. Plume Shape Observation and Prediction

In this research, the plume shape is defined by the plume depth, width and rise. Examination of the plume shape was started with identification of the plume depth where the significant lifting starts. Based upon filed observations, the plume depth is defined as following: when the plume lifting velocity, w, increased alone the "x" direction to a point where it started decreasing, the distance x at this point was defined as the plume depth. To quantify this depth, the first two sets of "w" data were sorted from the smallest to largest and each data was ranked. After data sorting and ranking it was discovered that, like what **Figure 12** has shown, the maximum of "w" appeared at the lower left part (x = 2.10 - 3.00 m, y = 0.00 - 2.14 m, z = 0.30 - 1.22 m) on the first ring, followed by points at the center (x = 8.3 m, y = 3.49 m, z = 1.22 m) on the second ring. On the third

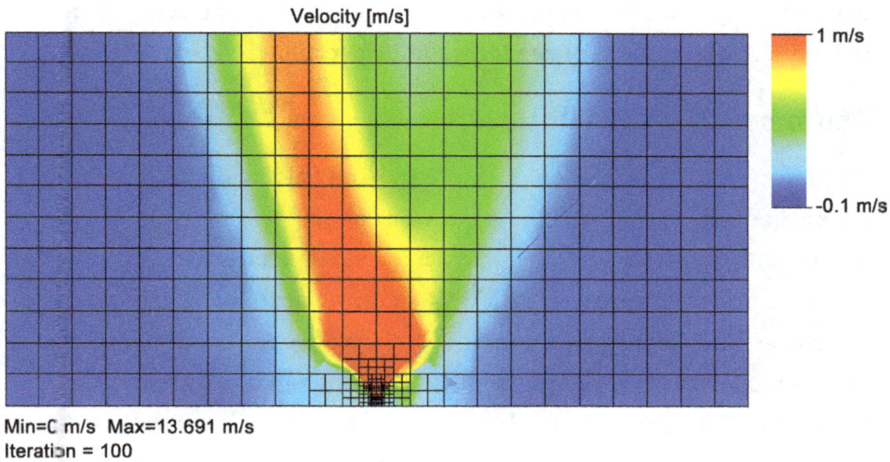

Figure 11. Top-view of CFD modeled contour plot of 3D velocities at 0.3 m above the ground.

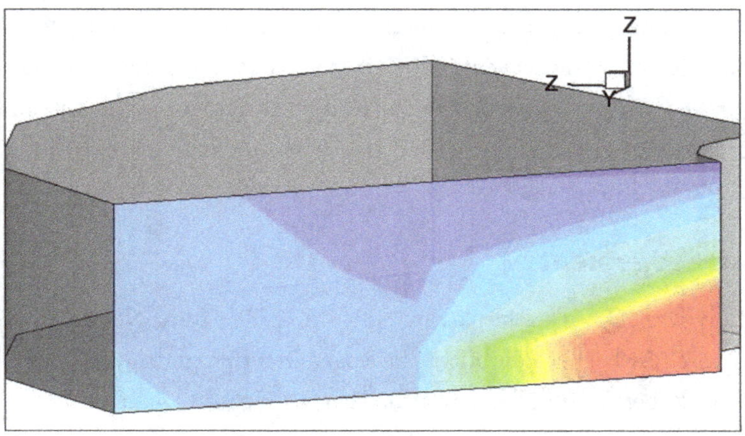

Figure 12. Side view of the contour plot of the velocity "w" modeled by the CFD.

ring, the biggest "w" appeared at the upper left part (x= 10.49 - 13.83 m, y = 5.81 - 10.72 m, z = 3.05 - 4.88 m), but these points had smaller w's as compared to the first two rings. This suggests that at 3 m away from the fan, there was a strong lifting momentum; at around 9 m away from the fan, the lifting momentum was still big, but not as strong as at the 3 m ring.

Combining the data examination with the plots visualization, it was discovered that the plume depth was approximately 9 m away from the fan where the plume started to lift.

Similar method was used to examine the plume width and it was discovered that the plume width was about ±6 m.

For plume rise examination, the filed measurement was conducted with the furthest distance at 15 m and top height at 4.88 m. The field smoke test indicated that the plume went beyond 15 m away from fan and rose over 4.88 m. Limitation of the measurement height and distance prevent a numerical development of the plume rise at this time. Further experiment is planned to extend field measurement scope. Moreover, further CFD modeling will be conducted with different fan operation conditions examine ventilation rate impact on the plume field, and to calibrate the CFD output such that this modeling approach may be used to predict full scale of plume lifting the dispersion.

4. Summary and Conclusion

In this research, a combination of theoretical and field study was conducted to measure and simulate the plum rise and shape of air flow from a ventilation fan. The theoretical modeling of the plume shape was conducted using a commercial Computational Fluid Dynamics (CFD) package named FloEFD; the field measurements of the plume field was conducted using five 3D ultrasonic anemometers to simultaneously measure the air flow in the plume at various locations. The TECPLOT package was used to visualize the plume flow field based upon anemometer measurements. While the plume shapes were found to be left-shifted by the CFD model and TECPLOT visualization, the magnitudes of the 3D wind velocities from field measurement were found to be significantly larger than those from CFD model. The plume field measurements indicated that the plume of a 0.6 m (24-inch) ventilation fan had a depth about 9 m, a width about ±6 m, and a rise (lifting) beyond the highest measurement point, 4.88 m (16 ft). Further filed data collection is needed to extend measurement height to capture the true plume lifting height.

Acknowledgements

This study was supported in part by NSF CAREER Award No. CBET-0954673 and USDA NIFA Higher Education Challenger Grant Program 2013-70003-20929. Help from Roberto Munillia, Steven Badawi, Di Hu and Qianfeng Li for testing setup construction and field data collection is also thankfully acknowledged.

References

[1] USDA/NASS (2005) Livestock and Poultry-Production and Value 2004 Summary. U.S. Department of Agriculture (USDA), National Agricultural Statistics Service (NASS), Washington, DC.

[2] Heber, A.J., Bogan, W., Ni, J.Q., Lim, T.T., Ramirez-Dorronsoro, J.C., Cortus, E.L., Diehl, C.A., Hanni, S.M., Xiao, C., Casey, K.D., Gooch, C.A., Jacobson, L.D., Kozel, J.A., Mitloehner, F.M., Ndgwa, P.M., Rpbarge, W.P., Wang, L. and Zhang, R (2008) The Nation Air Emisison Monitoring Study: Overview of Barn Sources. Presented at ILES VIII, Publisher, City.

[3] National Research Council (2003) Air Emissions from Animal Feeding Operations: Current Knowledge, Future Needs. The National Academies Press, Washington DC.

[4] Wang, K., Kilic, I., Li, Q.F., Wang-Li, L.J., Bogan, W.L. and Heber, A.J. (2010) National 386 Air Emissions Monitoring Study: Emissions Data from Two Tunnel-Ventilated Layer Houses in North Carolina-Site NC2B. Final Report. Purdue University, West Lafayette.

[5] Briggs, G.A. (1971) Plume Rise: A Recent Critical Review. *Nuclear Safety*, **12**, 15-24.

[6] Maes, G., Cosemans, G., Kretzschmar, J., Janssen, L. and Vantongerloo, J. (1995) Comparison of 6 Gaussian Dispersion Models Used for Regulatory Purposes in Different Countries of the Eu. *International Journal of Environment and Pollution*, **5**, 734-747.

[7] Singal, S.P., Gera, B.S. and Pahwa, D.R. (1994) Application of Sodar to Air Pollution Meteorology. *International Journal of Remote Sensing*, **15**, 427-441. https://doi.org/10.1080/01431169408954084

[8] Yadigaroglu, G., and Munera, H.A. (1987) Transport of Pollutants—Summary Review of Physical Dispersion Models. *Nuclear Technology*, **77**, 125-149. https://doi.org/10.13182/NT87-A33979

[9] Awasthi, S., Khare, M. and Gargava, P. (2006) General Plume Dispersion Model (GPDM) for Point Source Emission. *Environmental Modeling & Assessment*, **11**, 267-276. https://doi.org/10.1007/s10666-006-9041-y

[10] Hiscox, A.L., Miller, D.R., Holmén, B.A., Yang, W. and Wang, J. (2008) Near-Field Dust Exposure from Cotton Field Tilling and Harvesting. *Journal of Environmental Quality*, **37**, 551-556. https://doi.org/10.2134/jeq2006.0408

[11] Cooper, C.D. and Alley, F.C. (2002) Air Pollution Control: A Design Approach. 3rd Edition, Waveland Press, Prospect Heights.

[12] Holmes, N.S. and Morawska, L. (2006) A Review of Dispersion Modeling and Its Application to the Dispersion of Particles: An Overview of Different Dispersion Models Available. *Atmospheric Environment*, **40**, 5902-5928.

[13] Wendt, J.F. and Anderson, J.D. (2009) Basic Philosophy of CFD. In: *Computational Fluid Dynamics: An Introduction*, Spinger Verlag, Berlin, Heidelberg, 5-8.

[14] Alič, G., Širok, B. and Hočevar, M. (2010) Method for Modifying Axial Fan's Guard Grill and Its Impact on Operating Characteristics. *Forsch Ingenieurwes*, **74**, 87-98. https://doi.org/10.1007/s10010-010-0118-z

[15] Gates, R.S., Casey, K.D., Xin, H., Wheeler, E.E. and Simmons, J.D. (2004) Fan Assessment Numeration System (FANS) Design and Calibration Specifications. *Transactions of ASAE*, **47**, 1709-1715. https://doi.org/10.13031/2013.17613

[16] Wang, X., Chen, R. and Zhou, Y. (2011) The Flow Field Analysis and Optimization of Diagonal Flow Fan Based on CFD. *Advanced Materials Research*, **201-203**, 2657-2660. https://doi.org/10.4028/www.scientific.net/AMR.201-203.2657

[17] Quinn, A.D., Wilson, M., Reynolds, A.M., Couling, S.B. and Hoxey, R.P. (2001) Modeling the Dispersion of Aerial Pollutants from Agricultural Buildings—An Evaluation of Computational Fluid Dynamics (CFD). *Computers and Electronics in Agriculture*, **30**, 219-235.

[18] Tecplot, Inc. (2011) Tecplot 360 User's Manual. Tecplot, Inc., Bellevue.

[19] Kun, L., Zhe, W., Han, R. and Ren, Z. (2011) The Numerical Simulation of Directional Solidification Process Temperature Field for Large-Scale Frustum of a Cone Ingot. *Manufacturing Process Technology*, **189-193**, 1476-1481.

[20] Gong, Y., Zhang, Y., Li, H. and Wang, W. (2009) A Study on Airflow Field of Super-High Speed Pectination Grinding Wheel Based on PIV. *Advances in Abrasive Technology*, **389-390**, 332-337.

[21] Muramalla, K., Chen, Y. and Hechanova, A. (2005) Simulation and Optimization of Homogeneous Decomposition of Sulfur Trioxide Gas on a Catalytic Surface. *Heat Transfer Division of the American Society of Mechanical Engineers*, **376**, 201-207.

[22] Turner, D.B. (1994) Workbook of Atmospheric Dispersion Estimates: An Introduction to Dispersion Modeling. 2nd Edition, CRC Press, Boca Raton.

[23] Pasquill, F. (1961) The Estimation of the Dispersion of Windborne Material. *Meteorology Magazine*, **90**, 33-40.

[24] Gifford, F.A. (1976) Consequences of Effluent Releases. *Nuclear Safety*, **17**, 68-86.

[25] Linacre, E. and Geerts, B. (1999) Roughness Length. The University of Wyoming, 358 WY. http://www.das.uwyo.edu/~geerts/cwx/notes/chap14/roughness.html

3

Short-Term Disturbance Effects of Outdoor Education Stream Classes on Aquatic Macroinvertebrates

Jon P. Bossley[1,2], Peter C. Smiley Jr.[3]

[1]Environmental Science Graduate Program, The Ohio State University, Columbus, OH, USA
[2]School of Natural and Social Sciences, Mount Vernon Nazarene University, Mount Vernon, OH, USA
[3]USDA Agricultural Research Service, Soil Drainage Research Unit, Columbus, OH, USA
Email: jon.bossley@mvnu.edu

Abstract

Outdoor education programs frequently offer stream classes that teach students how to assess water quality based on the composition of aquatic macroinvertebrate communities. Repeatedly using the same site for stream classes can cause disturbance that could negatively impact aquatic macroinvertebrates. However, the impact of outdoor education stream classes on short term temporal trends of aquatic macroinvertebrates has not been evaluated. Our objective was to quantify whether outdoor education stream classes caused localized and short-term impacts on the aquatic macroinvertebrate community. We sampled aquatic macroinvertebrates over a five day period in May 2014 from an impacted riffle subjected to repeated substrate disturbance by outdoor education stream classes and an unimpacted riffle that was not subjected to stream classes within an agricultural stream in central Ohio. We did not observe a difference in macroinvertebrate community response variables between different time periods or among days within the impacted riffle as part of our within riffle analysis. We documented that macroinvertebrate abundance and dorsoventrally flattened clinger abundance was greater within the unimpacted riffle than the impacted riffle. Macroinvertebrate evenness was greater in the impacted riffle than the unimpacted riffle. In the impacted riffle percent clingers was greater on Monday than on Friday. Conversely, in the unimpacted riffle percent clingers was greater on Friday than on Monday. Our results indicated that outdoor education stream classes did not alter short term temporal trends of macroinvertebrate community structure within an impacted riffle, but the stream classes resulted in differences in community structure between an impacted and unimpacted riffle. Our results suggest that outdoor education centers should avoid repeatedly using the same site for their stream classes to prevent negatively impacting aquatic macroinvertebrates.

Keywords

Substrate Disruption, Trampling, Macroinvertebrate Community, Riffles

1. Introduction

Time and change are continuous processes in any ecosystem, but their influence is accelerated in highly dynamic ecosystems like streams. Disturbance plays a major role in determining the population, community, and ecosystem structure and dynamics of streams [1]. Disturbance has been defined as "any relatively discrete event in time that disrupts ecosystem, community, or population structure and changes resources, substrate availability, or the physical environment" [2]. While a press disturbance (e.g., logging, chemical pollutants within stream sediments) may affect lotic ecosystems continuously for months or years, pulse disturbances (e.g., spates) can periodically impact lotic ecosystems relatively quickly (*i.e.*, hours, days, or weeks) [3]. Chemical disturbances as a result of agricultural runoff or industrial waste and physical disturbances from floods, construction projects, and land-use changes have received much attention in the literature [1] [3]. A newer body of literature has begun to evaluate the influence of disturbance caused by human recreational activity within natural areas on the biota [4] [5] [6] [7] [8]. In addition to considering the disturbance impacts of ecotourism and outdoor recreation, it is important to recognize that outdoor education programs may also cause disturbance to the biota.

Outdoor education is a popular, hands-on educational method conducted at outdoor sites and aimed at fostering appreciation for nature among school children. Many camps, nature centers, and parks offer outdoor education stream classes that often involve a focus on assessing water quality based on the abundance and diversity of pollution sensitive aquatic macroinvertebrates [9]. Instructors of stream classes typically encourage students to move and flip rocks within the streambed in search of aquatic macroinvertebrates. Additionally, outdoor education organizations typically use the same stream and often the same site for their stream classes as a result of familiarity, ease of access, and safety considerations. Subsequently, over the course of a year the same stretch of stream may be subjected to substrate disruption by extensive student activity. The influence of outdoor education stream classes on stream macroinvertebrates has not been documented previously. Understanding the ecological responses of stream macroinvertebrates to human instream activity as a result of stream classes is needed to help outdoor education centers, nature centers, and parks sustainably manage the streams that they use for their educational programs [10] [11].

Insights to the potential effects of outdoor education stream classes on stream macroinvertebrates can be found in previous studies examining the influence of scientific collecting and outdoor recreational activities on stream macroinvertebrates. Falk [12] discussed the bias that researchers may introduce into a field study by collecting and vouchering specimens and argued that researchers in effect become a new top predator in the ecosystem. Three studies [13] [14] [15] examined the effects of outdoor recreation activities on aquatic macroinvertebrates and found evidence of reduced macroinvertebrate density [13], reduced macroinvertebrate abundance [14], reduced taxa richness [14], and increased catastrophic drift [15] in response to instream recreational activities by humans. These studies demonstrate that human activity may act as a pulse disturbance by causing substrate movement thereby inducing drift, injury, or death to stream macroinvertebrates. Additionally, the frequent, repetitive usage of the same stream site over weeks, months, or years by outdoor education programs may act as a press disturbance.

We examined the effects of daily disturbance caused by repetitive activity of students participating in outdoor education stream classes on stream macroinvertebrates within an agricultural stream in central Ohio. This study is one of four studies conducted at multiple spatial and temporal scales from 2013 to 2015 to quantify the influence of outdoor education stream classes on aquatic macroinvertebrates [16]. In this manuscript we document the results of the study designed to examine the localized, short-term disturbance effects of outdoor education stream classes on aquatic macroinvertebrates, and to our knowledge this is the first study to do so. Previous studies have found that aquatic macroinvertebrates can recover quickly following a physical disturbance in as little as three days [17] [18] [19]. As a result we believed an evaluation of the short term impacts of stream classes was needed to complement the findings of our longer term studies. Specifically, we investigated whether differences in the aquatic macroinvertebrate community occurred between time periods and among days over a five day period within a riffle repeatedly subjected to outdoor education stream classes. We also evaluated whether differences in the aquatic macroinvertebrate community occurred between a riffle repeatedly subjected to outdoor education stream classes and an unimpacted upstream riffle over a five day period. Our objective was to quantify whether outdoor education stream classes negatively impacted the aquatic macroinvertebrate community between time periods within a day (morning and afternoon) or among days. We predicted that the macroinvertebrate community will exhibit a negative response (*i.e.*, declines in abundance, richness, diversity, and changes in the abundance of select guilds) to daily student activity that increases as the week progresses within an impacted riffle subjected to repeated disturbance by outdoor education classes. We also predicted that the macroinvertebrate community in the impacted riffle will exhibit a negative response to repeated disturbance by outdoor education stream classes over a five day period, while the macroinvertebrate community in the unimpacted riffle will remain unchanged by student activity.

2. Methods

2.1. Study Area and Sampling Sites

Our study was conducted at two riffles within the portion of the upper Alum Creek that flows through the grounds of the Heartland Outdoor School in central Ohio. The upper Alum Creek is a fourth order stream at Heartland Outdoor School, which is an outdoor education center providing overnight/multi-day hands-on nature and environmental science programs for K-12 students. The impacted riffle (40°23'01.014"N, 82°52'37.870"W) is a 7 m long riffle that has been used as part of stream classes taught by the Heartland Outdoor School since 2010. The impacted riffle is characterized by a forested riparian buffer (riparian width > 30.5 m) with an agricultural field bordering the riparian buffer on the right bank. Alum Creek at this site is a C-4 riffle-pool stream [20] with moderate sinuosity (1.3), no constrictions, broad confinement ratio (>10), and substrate composed primarily of coarse gravel (D50 = 38% coarse gravel substrate). The impacted riffle is the focal point for students in the stream study program that receives an average of 6000 participants annually. Stream classes are held in the impacted riffle in the spring (April to June) and autumn (September to November). Students learn that Group 1 pollution-sensitive macroinvertebrates that serve as indicators of water quality prefer the highly oxygenated waters and heterogeneous substrate of the riffle [9] [21]. As a result of student exploration and collecting efforts the impacted riffle experiences greater foot traffic, greater incidence of rock-turning, and greater levels of substrate movement compared to other nearby channel units and the upstream unimpacted riffle [16]. We also sampled aquatic macroinvertebrates from an unimpacted riffle (40°23'02.662"N, 82°52'36.171"W) that is 7 m long and located 91 m upstream of the impacted riffle to enable a between-riffles comparison as part of our study. The unimpacted riffle exhibits riparian, geomorphic, and substrate characteristics similar to the impacted riffle, but it has never been used as part of Heartland Outdoor School's stream classes. The absence of stream classes within the unimpacted riffle results in it experiencing little or no human foot traffic, rock-flipping, and substrate movement [16].

We conducted our study from 19 May 2014 to 23 May 2014, which corresponds to the dates that Heartland Outdoor School receives the largest school groups (based on enrollment numbers) during its academic year. These dates represent the seventh week in which stream classes were offered in the spring and 447 students participated in 19 individual stream classes at the impacted riffle during the study. We anticipated that if student instream activity can induce short-term disturbance, it was most likely to be evident during this week. The warm weather in May is conducive for stream classes and, consequently, the stream class program annually experiences its greatest level of participation in May. Previous research findings [16] indicated that activity, abundance, and taxa richness of the resident macroinvertebrate community in this stream are also greatest during spring, making this season ideal for evaluating the influence of stream classes. The mean daily discharge levels at the USGS gauge located 5.6 km downstream of the impacted riffle during our study was 3.5 m^3/sec with a minimum of 2.2 m^3/sec and a maximum of 7.0 m^3/sec, indicating safe conditions for stream classes to proceed on schedule.

2.2. Within-Riffle Assessment

The objective for this assessment was to determine if the aquatic macroinvertebrate community within the impacted riffle differed between time periods (morning and afternoon) and among days during the week. Our sampling schedule was based on the Heartland Outdoor School's typical daily class schedule for the busy spring season (**Table 1**). Students typically arrive at the school on Monday morning and begin their slate of classes early in the afternoon. To ensure that all students have opportunity to take the stream class, the typical schedule rotation incorporates five stream classes per day. Two stream classes are conducted in the morning, three stream classes in the afternoon, and one stream class in the evening from Tuesday through Thursday. Friday stream classes are limited to the morning before the students return home. Following this schedule, morning samples for the within-riffle assessment were collected starting at 9:30 am EDT on Monday. From Tuesday to Thursday, morning samples were collected between 7:00-8:30 am EDT prior to the first stream classes for each day. Afternoon samples from Tuesday to Thursday were collected between 2:30-4:30 pm EDT after most afternoon stream classes were completed. The school group in session during the week of our study had an early departure time on Friday. Consequently, Friday morning samples for the within-riffle study were taken prior to 7:30 am EDT and the post-class samples were taken starting shortly after 11:00 am EDT. Within this sampling schedule our morning samples represented the condition of aquatic communities before daily disturbance (*i.e.*, prior to the first stream classes of the day) and our afternoon samples represented the condition of the aquatic macroinvertebrates after daily disturbance (*i.e.*, after stream classes were conducted).

A random sampling grid of 30 1 m^2 quadrats was set up over half of the riffle

Table 1. Typical stream class schedule used by Heartland Outdoor School from mid-April to early June of 2014 showing daily class times and a continuous count of stream classes for the week.

Class Time	Monday	Tuesday	Wednesday	Thursday	Friday
9:15-10:15 am		Class 4	Class 9	Class 14	Class 19
10:30-11:30 am		Class 5	Class 10	Class 15	Class 20
1:30-2:30 pm	Class 1	Class 6	Class 11	Class 16	
2:45-3:45 pm	Class 2	Class 7	Class 12	Class 17	
6:30-7:30 pm	Class 3	Class 8	Class 13	Class 18	

along the left bank where students enter the stream. If student-induced disturbance occurs, its effects were expected to be greater and more likely to be revealed in this half of the riffle because of the greater foot traffic. The random sampling grid ensured random selection of samples within the riffle, ensured representative sampling from three longitudinal sections (*i.e.*, upstream, middle, and downstream sections) of the riffle, and avoided repeat sampling of the same quadrat.

Aquatic macroinvertebrates were collected via Surber sampler (500 μm net; 30.5 cm × 30.5 cm frame) with the substrate disturbed by hand with the aid of a screwdriver for 30 seconds to a depth of 10 cm or to an impenetrable hard layer (whichever was encountered first). To ensure representative sampling from the entire riffle length, one sample was collected from a randomly determined quadrat within each of three sections of the riffle during each sampling period. The three riffle sections included upper riffle (*i.e.*, the two upstream rows in the sampling grid), mid riffle (*i.e.*, the two rows in the middle of the riffle), and lower riffle (*i.e.*, the two rows at the downstream end of the riffle). All three samples were collected consecutively beginning downstream and working upstream. Thus, we collected three randomly located Surber samples from the riffle each time period each day resulting in six samples collected each day and a total of 30 samples collected for the entire week.

Identification of live aquatic macroinvertebrates in each sample commenced immediately upon completion of sampling. Samples were individually transferred to a large picking pan and identified to phylum, class, order, or family taxonomic resolution with the aid of a Nikon (20×) dissection microscope. Organisms identified at phylum, class, order, or sub-order taxonomic resolution included non-insects (Collembola, Hydracarina, Nematoda, Oligochaeta, Turbellaria) and insects (Anisoptera, Zygoptera, Hemiptera, Plecoptera). Family-level identification was used for some non-insects (Asellidae, Cambaridae, Pleuroceridae, Sphaeridae) and most insect taxa (Psephenidae, Elmidae, Simuliidae, Ceratopogonidae, Chaoboridae, Chironomidae, Tipulidae, Baetidae, Caenidae, Heptageniidae, Isonychiidae, Leptophlebiidae, Helicopsychidae). All organisms from each sample were returned to the quadrat from which they were taken immediately after identification. The catch and release protocol was used to allow the macroinvertebrates an opportunity to recolonize the quadrat from which they were taken and safeguard against the investigators introducing additional disturbance by permanently removing specimens from the stream [12]. Our field identification and subsequent release of live animals prohibited the use of a finer scale taxonomic resolution to either genus or species level.

2.3. Between-Riffle Assessment

Our objective for this assessment was to determine if aquatic macroinvertebrate communities differed between the start of sampling on Monday and end of sampling on Friday (Monday and Friday) and between riffles (impacted riffle and unimpacted riffle). We also established a random sampling grid of 30 1 m^2 quadrats over half of the unimpacted riffle along the right stream bank. This portion of the unimpacted riffle was selected because it was physically similar to the impacted riffle. We sampled aquatic macroinvertebrates with a Surber sampler using the same sampling methods as described above within the unimpacted riffle. The quadrats within the unimpacted riffle that were selected for macroinvertebrate sampling were based on the quadrat numbers that were randomly selected for sampling within the impacted riffle. This ensured our samples were randomly selected and ensured comparability between the two riffles. The first samples from the unimpacted riffle were collected on Monday beginning at 11:45 am EDT and the last samples were collected on Friday afternoon starting at 12:00 pm EDT. Additionally, after identification all macroinvertebrates were released alive within the quadrat from which they were captured.

2.4. Statistical Analysis

We assessed macroinvertebrate community structure with nine response variables: 1) macroinvertebrate abundance (number of macroinvertebrates captured in each sample); 2) taxa richness (number of taxa represented in each sample); 3) EPT abundance (number of Ephemeroptera, Plecoptera, and Trichoptera captured in each sample); 4) clinger abundance (number of macroinvertebrates in the clinger habit guild including families Elmidae, Heptageniidae, Baetidae, Helicopsychidae, Glossosomatidae, Hydropsychidae, Philopotamidae, Polycentropodidae, Psephenidae, Simuliidae); 5) dorsoventrally flattened (DVF) clinger abundance (number of Heptageniidae and Psephenidae captured in each sample); 6) percent EPT (percent composition of EPT calculated as EPT abundance divided by macroinvertebrate abundance); 7) Shannon diversity [22]; 8) evenness (calculated as $E_{1/D}$) [23]; and 9) percent clingers (percent composition of clinger macroinvertebrates calculated as clinger abundance divided by macroinvertebrate abundance).

These nine response variables were chosen to encompass a range of community response variables that would ensure a robust evaluation of the riffle community. Macroinvertebrate abundance and taxa richness were included to examine whether stream classes can reduce the number of macroinvertebrates or eliminate certain taxa. Macroinvertebrate abundance and taxa richness are of immediate importance to outdoor education stream classes because greater abundance enables better opportunity for students to catch macroinvertebrates and greater taxa richness provides better opportunity for students to catch a variety of macroinvertebrate types. Shannon diversity and evenness were chosen on the basis of their robustness and wide usage as measures of diversity and evenness. EPT response variables were included because of the known sensitivity to disturbance exhibited by EPT taxa. The three clinger response variables (clinger abundance, percent clingers, and DVF clinger abundance) were selected because we anticipated that clingers would be tolerant of the substrate disturbance caused by stream classes because their dorsoventrally flattened bodies and tarsal claws would enable them to maintain position on the substrate with the occurrence of disturbance.

We used R statistical software [24] and a level of significance of 0.05 for all statistical analyses. For the within-riffle analysis, we used a two factor generalized linear model analysis with the quasi-Poisson distribution and the glm function [24] to determine if whole integer response variables based on count data (macroinvertebrate abundance, taxa richness, EPT abundance, clinger abundance, DVF clinger abundance) differed between time periods (*i.e.*, morning and afternoon) or among days during the week. We also conducted a two factor analysis of variance with the lm function [24] to determine if non-whole integer response variables (Shannon diversity, evenness, percent clingers, percent EPT) differed between time periods or among days within the impacted riffle. Examination of the normal q-q plots with the qqPlot function in the car package [25] for these four response variables confirmed that their distributions did not differ from the normal distribution.

For the between-riffles analysis, we used a two factor generalized linear model analysis with the quasi-Poisson distribution and the glm function [24] to determine if whole integer response variables based on count data (macroinvertebrate abundance, taxa richness, EPT abundance, clinger abundance, DVF clinger abundance) differed between days (*i.e.*, the start of sampling on Monday morning and the end of sampling on Friday afternoon) and sites (impacted and unimpacted riffles). We also used a two factor analysis of variance and the lm function [24] to determine if non-whole integer response variables (Shannon diversity, evenness, percent clingers, percent EPT) differed between days and sites. Examination of the normal q-q plots with the qqPlot function in the car package [25] for these four response variables confirmed that their distributions did not differ from the normal distribution.

3. Results

Within the impacted riffle we documented 25 taxa from 988 captures over five days of sampling. The three most abundant taxa captured were Nematoda, Chironomidae, and Baetidae (**Table 2**). For the within-riffle analysis, no differences ($P > 0.05$) in the nine macroinvertebrate response variables occurred between time periods or among days (**Table 3**). Additionally, none of the macroinvertebrate response variables exhibited an interaction effect of time period and day (**Table 3**).

As part of our between-riffles analysis we documented 11 taxa from 76 captures within the impacted riffle and 14 taxa from 324 captures within the unimpacted riffle (**Table 4**). The three most commonly occurring taxa in the impacted riffle were Nematoda, Chironomidae, and Baetidae, while the three most commonly occurring taxa in the unimpacted riffle were Chironomidae, Nematoda, and Oligochaeta (**Table 4**). We observed a difference ($P < 0.05$) in

Table 2. Abundance, percentage, and frequency of occurrences of each macroinvertebrate taxa captured within a riffle subjected to repeated disturbance by outdoor education stream classes in Alum Creek, Ohio from 19 May to 23 May 2014.

Taxa	Abundance	Percentage	Frequency
Nematoda (nematodes)	507	51.3	5
Chironomidae (non-biting midges)	160	16.2	5
Baetidae (small minnow mayflies)	112	11.3	5
Plecoptera (stoneflies)	49	5.0	5
Oligochaeta (segmented worms)	32	3.2	5
Heptageniidae (flatheaded mayflies)	29	2.9	5
Turbellaria (flat worms)	26	2.6	4
Psephenidae (water pennies)	16	1.6	5
Simuliidae (black flies)	16	1.6	5
Elmidae (riffle beetles)	11	1.1	5
Chaoboridae (phantom midges)	4	0.4	1
Hydracarina (water mites)	4	0.4	2
Caenidae (small squaregill mayflies)	3	0.3	3
Leptophlebiidae (pronggilled mayflies)	3	0.3	1
Zygoptera (damselflies)	3	0.3	3
Asellidae (sowbugs)	2	0.2	2
Cambaridae (crayfishes)	2	0.2	2
Tipulidae (crane flies)	2	0.2	1
Ceratopogonidae (biting midges)	1	0.1	1
Collembola (springtails)	1	0.1	1
Helicopsychidae (snailcase caddisflies)	1	0.1	1
Hemiptera (true bugs)	1	0.1	1
Pleuroceridae (Pleurocerid snails)	1	0.1	1
Sphaeridae (fingernail clams)	1	0.1	1
Trichoptera (caddisflies)	1	0.1	1

Table 3. P values from a two factor generalized linear model analysis and two factor analysis of variance conducted to determine if aquatic macroinvertebrate community response variables differed between time periods (morning and afternoon) or among days within a riffle subjected to repeated disturbance by outdoor education stream classes in Alum Creek, Ohio from 19 May to 23 May 2014.

Response variable	Time period	Day of week	Time × Day interaction effect
Macroinvertebrate abundance	0.771	0.327	0.146
Taxa richness	0.570	0.913	0.887
EPT abundance	0.603	0.166	0.339
Percent EPT	0.237	0.544	0.856
Shannon diversity	0.436	0.979	0.850
Evenness	0.442	0.985	0.522
Clinger abundance	0.899	0.152	0.223
Percent clingers	0.486	0.428	0.825
DVF clinger abundance	0.937	0.567	0.554

Table 4. Number (%) of each macroinvertebrate taxa captured within an impacted riffle subjected to repeated disturbance by outdoor education stream classes and an unimpacted riffle in Alum Creek, Ohio sampled at the start of sampling on Monday 19 May 2014 and the end of sampling on Friday 23 May 2014.

Taxa	Impacted	Unimpacted
Nematoda (nematodes)	21 (27.6)	101 (31.2)
Chironomidae (non-biting midges)	19 (25.0)	112 (34.6)
Baetidae (small minnow mayflies)	12 (15.8)	21 (6.5)
Plecoptera (stoneflies)	7 (9.2)	8 (2.5)
Heptageniidae (flatheaded mayflies)	3 (3.9)	19 (5.9)
Leptophlebiidae (pronggilled mayflies)	3 (3.9)	0 (0.0)
Oligochaeta (segmented worms)	3 (3.9)	42 (13.0)
Psephenidae (water pennies)	3 (3.9)	5 (1.5)
Simuliidae (black flies)	3 (3.9)	0 (0.0)
Tipulidae (crane flies)	1 (1.3)	0 (0.0)
Turbellaria (flat worms)	1 (1.3)	0 (0.0)
Anisoptera (dragonflies)	0 (0.0)	1 (0.3)
Caenidae (small squaregill mayflies)	0 (0.0)	1 (0.3)
Elmidae (riffle beetles)	0 (0.0)	9 (2.8)
Hydracarina (water mites)	0 (0.0)	1 (0.3)
Isonychiidae (brushlegged mayflies)	0 (0.0)	2 (0.6)
Pleuroceridae (Pleurocerid snails)	0 (0.0)	1 (0.3)
Zygoptera (damselflies)	0 (0.0)	1 (0.3)

macroinvertebrate abundance and DVF clinger abundance between riffles (**Table 5**). Macroinvertebrate abundance and DVF clinger abundance were greater at the unimpacted riffle than the impacted riffle (**Figure 1**). There was no difference ($P > 0.05$) between days and no significant interaction effect of day x site for macroinvertebrate abundance or DVF clinger abundance (**Table 5**). We also observed a difference ($P < 0.05$) in evenness between riffles (**Table 5**), with greater evenness within the impacted riffle than the unimpacted riffle (**Figure 1**). A significant day x site interaction effect occurred for percent clingers that indicated the effect of day was different between riffles (**Table 5**). The percent clingers within the impacted riffle was greater ($P < 0.05$) at the start of sampling on Monday morning than at the end of sampling on Friday afternoon, while the percent clingers in the unimpacted riffle was greater ($P < 0.05$) at the end of sampling on Friday afternoon than the beginning of sampling on Monday morning (**Figure 2**). No difference ($P > 0.05$) in taxa richness, EPT abundance, percent EPT, Shannon diversity, or clinger abundance occurred between days or between sites (**Table 5**).

Table 5. P values from a two factor generalized linear model analysis and two factor analysis of variance conducted to determine if aquatic macroinvertebrate community response variables differed between days (Monday and Friday) and sites (a riffle impacted by repeated disturbance from outdoor education stream classes and an unimpacted riffle) within Alum Creek, Ohio on 19 May and 23 May 2014. Bolded P values are those that were <0.05.

Response Variable	Day	Site	Day × Site Interaction Effect
Macroinvertebrate abundance	0.914	**<0.001**	0.877
Taxa richness	0.460	0.107	0.539
EPT abundance	0.242	0.141	0.628
Percent EPT	0.289	0.050	0.141
Shannon diversity	0.549	0.932	0.285
Evenness	0.772	**<0.001**	0.730
Clinger abundance	0.552	0.080	0.237
Percent clingers	0.934	0.178	**0.039**
DVF clinger abundance	0.317	**0.039**	0.258

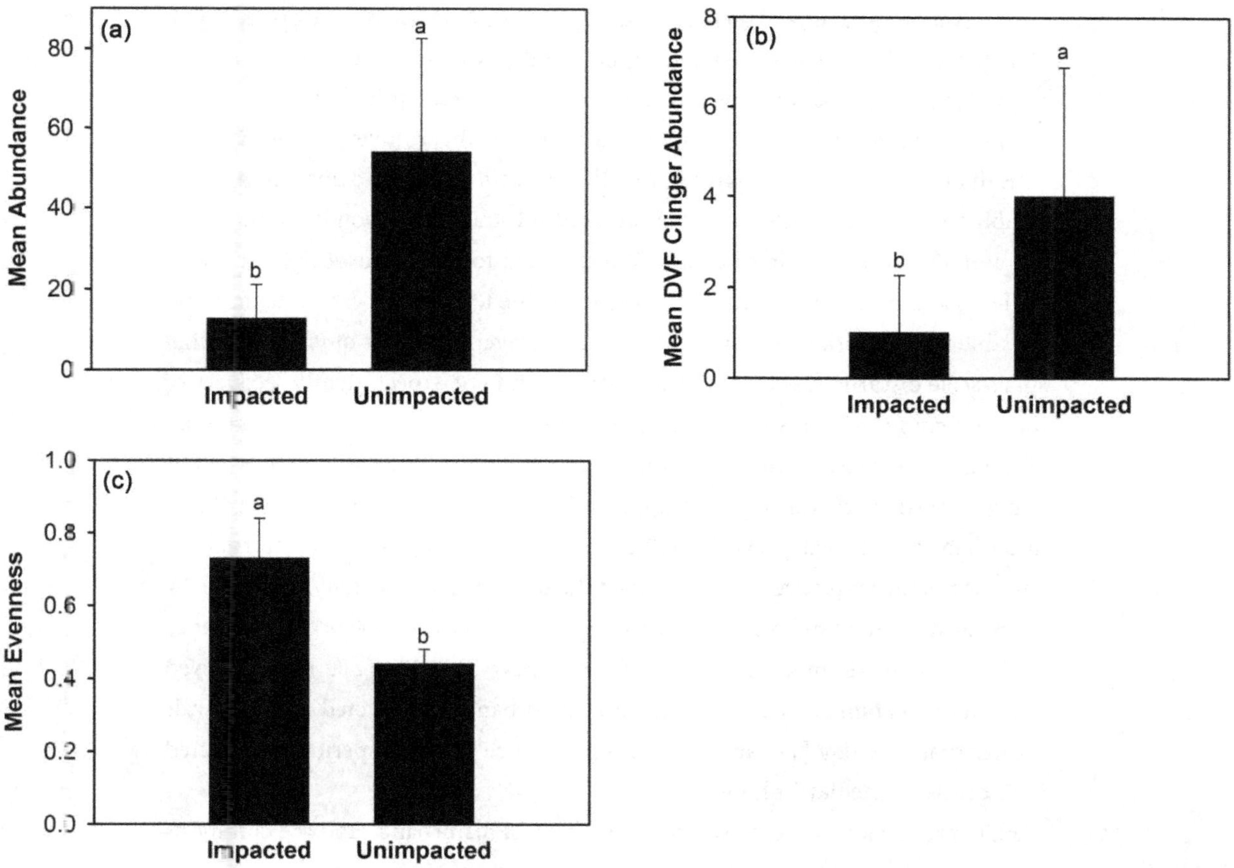

Figure 1. Mean macroinvertebrate abundance (a), mean dorsoventrally flattened (DVF) clinger abundance (b), and mean evenness ($E_{1/D}$, (c)) within an impacted riffle disturbed by outdoor education stream classes and an unimpacted riffle in Alum Creek, Ohio 19 May and 23 May 2014.

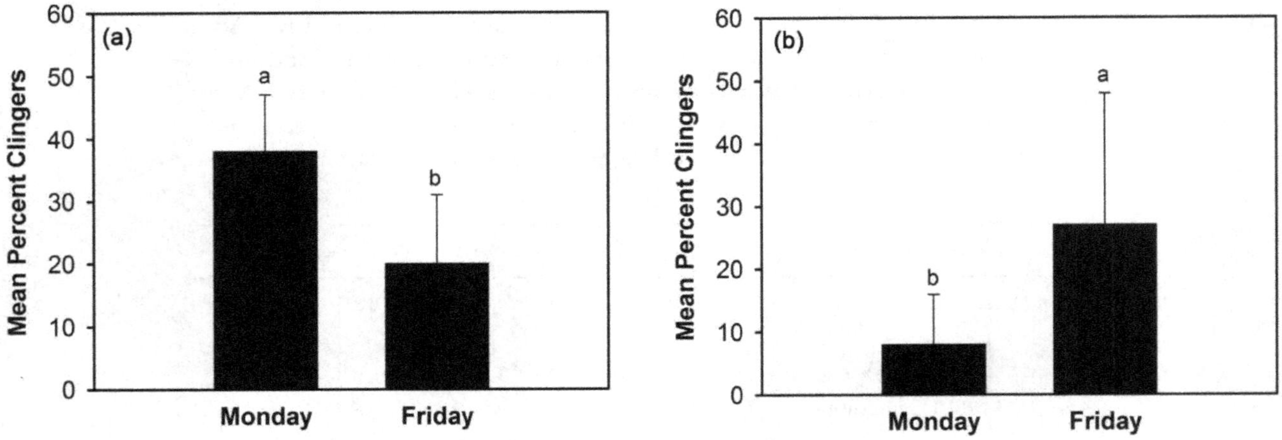

Figure 2. Mean percent clingers at the start of sampling Monday 19 May 2014 and the end of sampling on Friday 23 May 2014 within an impacted riffle disturbed by outdoor education stream classes (a) and an unimpacted riffle (b) in Alum Creek, Ohio.

4. Discussion

4.1. Within-Riffle Assessment

If student-induced disturbance occurred at the impacted riffle, we predicted that we would observe declines in macroinvertebrate abundance, richness, and diversity, as well as changes in the abundance of select guilds between the morning and afternoon samples and an increased negative response as the week progressed. This pattern did not manifest during our study, which suggests that intensive stream exploration activity by students in the impacted riffle during our study did not result in disturbance to the resident aquatic macroinvertebrates. Our results did not support the hypothesis that the macroinvertebrate community would exhibit a negative response to student-induced substrate disruption between morning and afternoon and among days. We expected to observe overnight recovery in the macroinvertebrate community each day due to macroinvertebrate drift and the ability of aquatic macroinvertebrates to recover relatively quickly following small-scale disturbances [3] [15]. However, we did not expect nightly recovery of the nine community response variables to reach the levels exhibited at the beginning of the study by each morning. The drift phenomenon is known to occur at night, particularly during dusk, apparently to avoid diurnal predators that rely upon light to see their prey [26] [27]. However, no clear pattern of diurnal decline or nocturnal recovery was observed during our five day study. The absence of change between time periods and among days in macroinvertebrate abundance or taxa richness within the impacted riffle contrasts with studies that have observed reductions in abundance and richness for disturbances conducted over intervals greater than one day [17] and for post-disturbance sampling periods conducted in intervals of one day or longer [18].

EPT taxa are known to be sensitive to chemical disturbance and are commonly used as water quality indicators in stream monitoring [21]. Since student-induced rock movement represents a potential physical disturbance rather than a chemical disturbance, it is not surprising that EPT-related response variables did not differ between time periods or among days during our study. In contrast, Laing [14] observed reduced EPT abundance during the summer (June to August) in tributaries of the Niobrara River in northern Nebraska disturbed by humans walking across the substrate when compared with similar undisturbed tributaries. However, Laing [14] did not observe a difference in EPT abundance between disturbed and undisturbed tributaries in May. This similarity between our results and Laing's [14] May results suggest that EPT responses to human activity may differ among seasons. Furthermore, EPT response to physical disturbance may relate more to the habit guild to which the taxa belong, rather than their taxonomic classification. Most Plecoptera and many Ephemeroptera are sprawlers and crawlers that may be more susceptible to physical disturbance than clingers. If the relative abundance of EPT sprawlers and crawlers is greater in summer months compared to EPT clingers, this may account for their seasonal responsiveness to physical disturbance.

Physical disturbance in the form of substrate rearrangement during spates is known to cause a decline in macroinvertebrate density and/or taxa richness [28] [29] [30] [31] [32]. However, spates not only cause movement of streambed rocks, they also increase the water velocity and discharge. Macroinvertebrates that get dislodged from the streambed during substrate movement in a spate are prone to being swept downstream by the increased stream power. This dual disturbance of substrate movement and increased water velocity does not occur as a result of student exploration of streams during outdoor education stream classes. Macroinvertebrates dislodged from the substrate during student-induced substrate disturbance may simply crawl or swim to another rock, drift to a neighboring patch, or retreat into the hyporheic zone. For example, Baetidae nymphs are quick, agile swimmers and have been observed to dart quickly from one patch to another when the stone under which they reside is removed (J. Bossley, personal observation). Macroinvertebrates that typically occupy the hyporheic zone may simply retreat further into the hyporheic zone upon sensing vibrations from students hiking on the streambed [31] or upon sensing a change in water velocity caused by rearrangement of rocks.

We evaluated three clinger response variables (clinger abundance, percent clingers, and DVF clinger abundance) because of their anticipated tolerance to physical disruption. Clingers that exhibit dorsoventral flattening of the body and/or possess claws that enable proficient clinging to the substrate may be able to resist being dislodged from a rock when it is moved or picked up out of the stream by a student. Clingers like Heptageniidae nymphs have been observed scurrying from one side of a rock to the other as it is inspected and turned (J. Bossley, personal observation). Slow moving clingers like Psephenidae larvae may simply remain in place if not discovered and removed from the rock. However, our results did not indicate whether clingers were more tolerant to disturbance caused by stream classes compared to other habit guilds.

An important consideration in interpreting the results of our within-riffle assessment is the catch and release policy that Heartland Outdoor School maintains for their stream classes. All organisms captured during each stream class are released back into the water at the left edge of the stream at the impacted riffle. Although some macroinvertebrates inevitably incur injury or death during each class, most are safely released back into the riffle (J. Bossley, personal observation), where they have opportunity to recolonize before the next class begins. This observation contrasts with Clifford's [33] report that most organisms captured after experimental disturbance of streambed patches were severely damaged or dead. However, the substrate in Clifford's [34] stream consisted of fine particles (mean width = 1.7 mm) and was vigorously disturbed with rakes, hoes, and shuffling of feet down to the hardpan. In Alum Creek the substrate is primarily coarse gravel (width = 32 - 64 mm) with numerous cobbles affording greater refugia via greater rock surface area and larger pore spaces in the gravel. Also, the student-induced substrate disturbance and macroinvertebrate collection methods used as part of Heartland Outdoor School's stream classes is likely less disruptive to the substrate than the experimental substrate disturbance implemented as part of Clifford's [33] study.

Since most captured macroinvertebrates are safely released back into the stream, they have opportunity to redistribute throughout the riffle. Typically, the interim period from the time one class leaves the stream until the next class enters the stream is 30 to 45 minutes. This is likely a sufficient period for swimming organisms such as Cambaridae, Odonata, and Baetidae to disperse into the stream and find a new suitable location. It has been reported that some *Baetis* spp. are capable of reaching post-disturbance abundance levels on experimental substrate within one day [17]. Even some Tipulidae, while burrowers by nature, possess adequate swimming ability to disperse into a more suitable part of the stream. Slower moving crawlers and clingers such as Heptageniidae, Leptophlebiidae, Psephenidae, and Elmidae larvae are likely to drift downstream upon release or find a nearby rock on which to attach. Burrowing or clinging organisms such as Chironomidae and Simuliidae are also likely to either drift downstream, return to the hyporheic zone, or attach to a rock near the release point.

Students moving throughout the impacted riffle, overturning rocks, and moving rocks during stream classes resulted in the substrate within the impacted riffle being disturbed in a patchy manner [16]. Since the impacted riffle consists of similar substrate throughout the entire riffle, only fine-grained differences in substrate disturbance occurs among patches within the riffle. Released macroinvertebrates, drifting macroinvertebrates from upstream, and macroinvertebrates able to avoid capture are free to continuously redistribute [34] throughout patches within the impacted riffle, limited only by their own size and mobility and whatever dispersal barriers those may convey [35].

Future research should examine whether the composition of habit guilds changes among different patches within the impacted riffle during the course of a day or week as classes continue to remove and release organisms because of the potential cumulative disturbance effect of classes throughout the day. It is possible that the frequent catch and release of macroinvertebrates by students occurs at a faster rate than the macroinvertebrates can redistribute themselves. We did not document a decline in clingers within the impacted riffle during the study, but this could reflect clingers' ability to avoid detection or remain attached even if their rocks were removed from the stream and examined. If clingers were difficult to see and collect from rocks at the start of the week they would still be equally difficult to see and collect at the end of the week, resulting in no cumulative decline throughout the week. This may be especially true for Psephenidae larvae, since they are DVF clingers that are often well-camouflaged against sandstone rocks as they typically remain nearly or entirely still when the rock to which they are attached is temporarily removed from the stream by a student. Sprawlers and crawlers such as Plecoptera are more likely to be seen and removed from a rock when inspected by a student because of their tendency to move in response to the rock being removed from the water. Also, crawlers are more likely to be knocked loose from rocks during disturbance of the substrate when a kick seine is used because of their reduced clinging ability compared to clingers. Consequently, a disturbance regime induced by students may lead to a shift in sprawlers and crawlers throughout the day or week from the right and center sides of the riffle where they are collected toward the left side of the riffle where these organisms are released after each class. Alternatively, the tendency for non-movement and DVF body form that Psephenidae exhibit may not be an advantage to them in avoiding detection by students, so a student-induced disturbance regime may also lead to an increase in clingers and DVF clingers on the left side of the riffle.

4.2. Between-Riffle Assessment

For the between-riffles assessment, we predicted that no significant change in macroinvertebrate community structure would occur from the start of sampling on Monday morning to the end of sampling on Friday afternoon at the unimpacted riffle, while a cumulative decline in the macroinvertebrate community structure was expected from the start of sampling on Monday morning to the end of sampling on Friday afternoon within the impacted riffle. Overall, our results did not support the hypothesis that the macroinvertebrate community in the impacted riffle would exhibit a negative response to student-induced substrate disruption from the start of sampling on Monday morning to the end of sampling on Friday afternoon, while the macroinvertebrate community in the unimpacted riffle remained unchanged.

Our results indicated that macroinvertebrate abundance was greater in the unimpacted riffle than the impacted riffle at the start of sampling on Monday morning before stream classes began and the end of sampling on Friday afternoon after stream classes ended for the week and suggests that student-induced disturbance as a result of outdoor education stream classes led to the observed abundance trends. Our finding that student-induced disturbance decreased macroinvertebrate abundance is consistent with the findings of other studies [13]

[14] [15] that documented decreased macroinvertebrate abundance, density, and increased macroinvertebrate drift in response to human recreational activity. However, Caires *et al.* [15] also documented that benthic macroinvertebrate abundance did not differ among riffles subjected to different levels of disturbance by hikers. Caires *et al.* [15] explained their contradictory results by concluding that instream hiking increased macroinvertebrate drift, but it did not result in long-term impacts on macroinvertebrates because individuals from nearby undisturbed patches were able to recolonize the disturbed patches.

The observed difference in evenness between the riffles suggests that student-induced substrate disruption influenced the pattern of relative abundance of macroinvertebrate taxa within the impacted riffle. The relative abundance of larval Chironomidae, one of the top two most abundant taxa in both riffles, was ten percent greater in the unimpacted riffle than in the impacted riffle. The decreased relative abundance of Chironomidae within the impacted riffle concurs with the findings of Gafner and Robinson [36] who documented decreases in larval Chironomidae relative abundance with increasing disturbance within first order alpine streams in Switzerland. In contrast, larval Chironomidae relative abundance in streams within the United States typically increases within increasing disturbance [37] [38] [39]. Chironomidae larvae inhabiting upper Alum Creek during the spring are found most abundantly among the algae on rock surfaces (J. Bossley, personal observation). Perhaps the frequent student-induced substrate disturbance within the impacted riffle reduced Chironomid algal food sources and subsequently resulted in reduced abundance of Chironomidae larvae [16].

The greater DVF clinger abundance observed at the unimpacted riffle compared to the impacted riffle may be a reflection of the greater overall macroinvertebrate abundance at the unimpacted riffle. However, the difference ($P < 0.05$) in the temporal trends of percent clingers between riffles suggests that the difference in DVF clinger abundance between riffles is not simply a result of the difference in macroinvertebrate abundance between the riffles. The reduction in percent clingers from the start of sampling on Monday morning to the end of sampling on Friday afternoon within the impacted riffle is likely a result of the clingers' inability to reestablish at a faster rate than the substrate disruption occurs. The increase in clinger percentage from the beginning of sampling on Monday morning to the end of sampling on Friday afternoon within the unimpacted riffle may be attributed to the impact of sampling macroinvertebrates on Monday morning and the lack of student disruption of substrate throughout the rest of the week. Localized substrate disturbance within sampling quadrats on Monday morning in the unimpacted riffle may have opened these areas for colonization. Clingers from neighboring patches as well as drifters from upstream had sufficient time to recolonize the sampled plots by the end of the week and were able to do so without the interruption of student-induced substrate disturbance, which could have resulted in the observed increase in percent clingers within the unimpacted riffle.

Clingers, particularly DVF clingers, are better-suited to resist physical substrate disruption. Consequently, we anticipated that if aquatic macroinvertebrates exhibit differential response to student-induced disturbance based on habit guild, it may be revealed by a greater presence of clingers at the impacted riffle by the end of the week. However, the opposite trend occurred (**Figure 2**) within our results. This finding suggests that either rocks in the impacted riffle were thoroughly examined and picked by students such that clingers were unable to remain unseen and were as likely to be removed from rocks as other macroinvertebrates or the level of student-induced substrate disturbance throughout the week was enough to dislodge clingers and cause them to drift downstream. Clingers also appear to have been slower at recolonizing the impacted riffle than macroinvertebrates in other habit guilds, perhaps due to their reduced mobility and lesser ability to move upstream compared to crawlers and swimmers.

The differences in the macroinvertebrate community between the impacted riffle and unimpacted riffle may be due to the prolonged effect of student-induced rock movement at the impacted riffle. Heartland Outdoor School began using the impacted riffle for stream classes in 2010. Consequently, at the time this study was conducted, the impacted riffle had received five years of continuous use in the spring and autumn by outdoor education stream classes. We have also quantified that riffles subjected to outdoor education stream classes experience increased substrate disturbance by the students and subsequently reduced macroinvertebrate colonization compared to undisturbed riffles [16]. Thus, the observed difference in macroinvertebrate abundance, evenness, percent clingers, and DVF clinger abundance between impacted and unimpacted riffles likely reflects the frequent substrate disruption at the impacted riffle over time. Matthaei *et al.* [18] postulated that disturbance frequency may result in biotic consequences only if it persists over long durations (*i.e.*, >30 days). Viewed on a seasonal or annual time scale, the repeated student-induced substrate disruption at the impacted riffle may have acted as a press disturbance.

Alternatively, the differences between riffles in macroinvertebrate abundance, evenness, percent clingers, and DVF clinger abundance may indicate that the unimpacted riffle consists of more suitable habitat allowing it to host a macroinvertebrate community containing more individuals, reduced evenness in relative abundance among different taxa, and greater numbers of clingers and DVF clingers. We did not measure water velocity and discharge during this study. However, both riffles were selected because they were qualitatively similar in wet width, amount of turbulent flow, degree of water surface broken by underlying substrate, and water depth. Our site selection criteria and the close proximity of the two riffles to each other (*i.e.*, within 100 m) suggests that hydrologic differences between riffles are not likely to result in the observed differences in macroinvertebrates. Additionally, visual assessment of the substrate at both riffles indicated similar substrate composition, which suggests that substrate type is not likely a potential cause for the observed difference in macroinvertebrate communities between riffles. We did not measure water chemistry during our study. However, EPT abundance and percent EPT did not differ between the impacted and unimpacted riffles. The sensitivity of EPT taxa to water chemistry conditions suggests that water chemistry conditions were similar at both riffles and likely did not lead to the observed differences in macroinvertebrate community structure between the impacted and unimpacted riffles.

No sampling was conducted in either riffle prior to 2010 when the stream class program was initiated, so information on the macroinvertebrate community within the riffles prior to student impact is not available. However, apparent similarities in hydrologic conditions, substrate type, and water chemistry between the unimpacted riffle and the impacted riffle suggest that differences in macroinvertebrate abundance, DVF clinger abundance, percent clingers, and evenness are due to prolonged effects of student-induced substrate rearrangement at the impacted riffle. Further study is needed to confirm these observations.

5. Conclusion

In conclusion, the absence of any clear decline and recovery pattern among the macroinvertebrate community during our study indicates that outdoor education stream classes did not lead to changes in aquatic macroinvertebrate community structure between time periods and among days at the within-riffle scale. We believe the lack of differences in community structure at the within-riffle scale resulted from the continuous redistribution of macroinvertebrates within and among patches within the impacted riffle. Over a longer time scale and at the between-riffle scale, our results indicate that the cumulative effects of outdoor education stream classes may have acted as a press disturbance and negatively impacted the biota within the impacted riffle. Our findings suggest that directors and administrative staff of outdoor education centers should be aware that their stream classes may negatively impact the macroinvertebrate communities. Directors and administrative staff of outdoor education centers should consider using multiple riffle sites in an alternating fashion for their educational efforts instead of using one riffle exclusively for stream classes to reduce the potential impacts of stream classes on aquatic macroinvertebrates.

Acknowledgements

We thank S. Fisher, R. Moore, K. Lekies, D. Denlinger, and an anonymous reviewer for helpful comments on an earlier draft of this manuscript. We also thank program manager A. Buzbee of Heartland Outdoor School for supporting the project and permission to access sites on the Heartland Outdoor School property for sampling. We thank H. Nguyen for providing guidance on statistical analyses. JPB was supported by a National Science Foundation Fellowship provided by the grant Linking Watershed Education with GK-12 Education in a Watershed Context.

References

[1] Stanley, E.H., Powers, S.M. and Lottig, N.R. (2010) The Evolving Legacy of Disturbance in Stream Ecology: Concepts, Contributions, and Coming Challenges. *Journal of the North American Benthological Society*, **29**, 67-83. https://doi.org/10.1899/08-027.1

[2] Pickett, S.T.A. and White, P.S. (1985) The Ecology of Natural Disturbance and Patch Dynamics. Academic Press, Orlando.

[3] Niemi, G.J., Devore, P., Detenbeck, N., Taylor, D., Lima, A., Pastor, J., Yount, J.D. and Naiman, R.J. (1990) Overview of Case Studies on Recovery of Aquatic Systems from Disturbance. *Environmental Management*, **14**, 571-587. https://doi.org/10.1007/BF02394710

[4] Davis, A.K. (2007) Walking Trails in a Nature Preserve Alter Terrestrial Salamander Distributions. *Natural Areas Journal*, **27**, 385-389. https://doi.org/10.3375/0885-8608(2007)27[385:WTIANP]2.0.CO;2

[5] Smith-Castro, J.R. and Rodewald, A.D. (2010) Effects of Recreational Trails on Northern Cardinals (*Cardinalis cardinalis*) in Forested Urban Parks. *Natural Areas Journal*, **30**, 328-337. https://doi.org/10.3375/043.030.0308

[6] Weston, M.A., Schlacher, T.A. and Lynn, D. (2014) Pro-Environmental Beach Driving Is Uncommon and Ineffective in Reducing Disturbance to Beach-Dwelling Birds. *Environmental Management*, **53**, 999-1004. https://doi.org/10.1007/s00267-014-0256-4

[7] Santos, G.S., Burgos, D.C., Lira, S.M.A. and Schwamborn, R. (2015) The Impact of Trampling on Reef Macrobenthos in Northeastern Brazil: How Effective Are Current Conservation Strategies? *Environmental Management*, **56**, 847-858. https://doi.org/10.1007/s00267-015-0552-7

[8] Guo T., Smith, J.W., Leung, Y.F., Seekamp, E. and Moore, R.L. (2015) Determinants of Responsible Hiking Behavior: Results from a Stated Choice Experiment. *Environmental Management*, **56**, 765-776. https://doi.org/10.1007/s00267-015-0513-1

[9] Firehock, K., Marshall, L. and Pinto, I. (1995) Hands on Save Our Streams: The Save Our Streams Teacher's Manual. Izaak Walton League of America, Gaithersburg.

[10] Barrows, C.W. and Allen, M.F. (2007) Biological Monitoring and Bridging the Gap between Land Management and Science. *Natural Areas Journal*, **27**, 194-197. https://doi.org/10.3375/0885-8608(2007)27[194:BMABTG]2.0.CO;2

[11] Willson, G.D., Narumalani, S., Lockert, C.K. and Merani, P.B.T. (2014) Application of Natural Resource Condition Assessments to Two Small National Parks. *Natural Areas Journal*, **34**, 235-248. https://doi.org/10.3375/043.034.0214

[12] Falk, J.H. (1974) Estimating Experimenter-Induced Bias in Field Studies: A Cautionary Tale. *Oikos*, **25**, 374-378. https://doi.org/10.2307/3543959

[13] Wright, K.K. and Li, J.L. (1998) Effects of Recreational Activities on the Distribution of *Dicosmoecus gilvipes* in a Mountain Stream. *Journal of the North American Benthological Society*, **17**, 535-543. https://doi.org/10.2307/1468370

[14] Laing, K.D. (2008) Impacts of Human Physical Disturbance on Benthic Communities in Tributaries of the Niobrara National Scenic River. MS Thesis, University of Nebraska, Lincoln.

[15] Caires, A.M., Vinson, M.R. and Brasher, A.M.D. (2010) Impacts of Hikers on Aquatic Macroinvertebrates in the North Fork of the Virgin River, Utah. *Southwestern Naturalist*, **55**, 551-557. https://doi.org/10.1894/JS-33.1

[16] Bossley, J.P. (2016) Environmental Impact from Outdoor/Environmental Education Programs: Effects of Frequent Stream Classes on Aquatic Macroinvertebrates. PhD Dissertation, Ohio State University, Columbus.

[17] Robinson, C.T. and Minshall, G.W. (1986) Effects of Disturbance Frequency on Stream Benthic Community Structure in Relation to Canopy Cover and Season. *Journal of the North American Benthological Society*, **5**, 237-248. https://doi.org/10.2307/1467711

[18] Matthaei, C.D., Uehlinger, U., Meyer, E.I. and Frutiger, A. (1996) Recolonization by Benthic Invertebrates after Experimental Disturbance in a Swiss Prealpine River. *Freshwater Biology*, **35**, 233-248. https://doi.org/10.1046/j.1365-2427.1996.00496.x

[19] Dole-Olivier, M.J., Marmonier, P. and Beffy, J.L. (1997) Response of Invertebrates to Lotic Disturbance: Is the Hyporheic Zone a Patch Refugium? *Freshwater Biology*, **37**, 257-276. https://doi.org/10.1046/j.1365-2427.1997.00140.x

[20] Rosgen, D. (1996) Applied River Morphology. Wildland Hydrology. Pagosa Springs.

[21] US EPA (1997) Volunteer Stream Monitoring: A Methods Manual. Report 841-B-97-003, U.S. Environmental Protection Agency, Office of Water, Washington DC.

[22] Magurran, A.E. (2004) Measuring Biological Diversity. Blackwell Publishing, Malden.

[23] Smith, B. and Wilson, J.B. (1996) A Consumer's Guide to Evenness Indices. *Oikos*, **76**, 70-82. https://doi.org/10.2307/3545749

[24] R Core Team (2015) R (Version 3.2.3): A Language and Environment for Statistical Computing. R Foundation for Statistical Computing, Vienna.

[25] Fox, J. and Weisberg, S. (2011) An R Companion to Applied Regression. Sage Publishing, Thousand Oaks.

[26] Williams, D.D. and Hynes, H.B.N. (1976) The Recolonization Mechanisms of Stream Benthos. *Oikos*, **27**, 265-272. https://doi.org/10.2307/3543905

[27] Sagar, P.M. and Glova, G.J. (1992) Invertebrate Drift in a Large, Braided New Zealand River. *Freshwater Biology*, **27**, 405-416. https://doi.org/10.1111/j.1365-2427.1992.tb00550.x

[28] Cobb, D.G., Galloway, T.D. and Flanagan, J.F. (1992) Effects of Discharge and Substrate Stability on Density and Species Composition of Stream Insects. *Canadian Journal of Fisheries and Aquatic Sciences*, **49**, 1788-1795. https://doi.org/10.1139/f92-198

[29] Boulton, A.J., Peterson, C.G., Grimm, N.B. and Fisher, S.G. (1992) Stability of an Aquatic Macroinvertebrate Community in a Multiyear Hydrologic Disturbance Regime. *Ecology*, **73**, 2192-2207. https://doi.org/10.2307/1941467

[30] Downes, B.J., Lake, P.S., Glaister, A. and Webb, J.A. (1998) Scales and Frequencies of Disturbances: Rock Size, Bed Packing and Variation among Upland Streams. *Freshwater Biology*, **40**, 625-639. https://doi.org/10.1046/j.1365-2427.1998.00360.x

[31] Matthaei, C.D., Arbuckle, C.J. and Townsend, C.R. (2000) Stable Surface Stones as Refugia for Invertebrates During Disturbance in a New Zealand Stream. *Journal of the North American Benthological Society*, **19**, 82-93. https://doi.org/10.2307/1468283

[32] Death, R.G. (2003) Spatial Patterns in Lotic Invertebrate Community Composition: Is Substrate Disturbance Actually Important? *Canadian Journal of Fisheries and Aquatic Sciences*, **60**, 603-611. https://doi.org/10.1139/f03-052

[33] Clifford, H.F. (1982) Effects of Periodically Disturbing a Small Area of Substratum

in a Brown-Water Stream of Alberta, Canada. *Freshwater Invertebrate Biology*, **1**, 39-47. https://doi.org/10.2307/3259453

[34] Peckarsky, B.L. (1986) Colonization of Natural Substrates by Stream Benthos. *Canadian Journal of Fisheries and Aquatic Sciences*, **43**, 700-709. https://doi.org/10.1139/f86-085

[35] Gore, J.A. and Milner, A.M. (1990) Island Biogeographical Theory: Can It Be Used to Predict Lotic Recovery Rates? *Environmental Management*, **14**, 737-753. https://doi.org/10.1007/BF02394722

[36] Gafner, K. and Robinson, C.T. (2007) Nutrient Enrichment Influences the Responses of Stream Macroinvertebrates to Disturbance. *Journal of the North American Benthological Society*, **26**, 92-102. https://doi.org/10.1899/0887-3593(2007)26[92:NEITRO]2.0.CO;2

[37] Barbour, M.T., Gerritsen, J., Snyder, B.D. and Stribling, J.B. (1999) Rapid Bioassessment Protocols for Use in Streams and Wadeable Rivers: Periphyton, Benthic Macroinvertebrates and Fish. Second Edition, EPA 841-B-99-002, U.S. Environmental Protection Agency, Washington DC.

[38] Hutchens Jr., J.J., Schuldt, J.A., Richards, C., Johnson, L.B., Host, G.E. and Breneman, D.H. (2009) Multi-Scale Mechanistic Indicators of Midwestern USA Stream Macroinvertebrates. *Ecological Indicators*, **9**, 1138-1150.

[39] Smiley Jr., P.C., King, K.W. and Fausey, N.R. (2010) Public Health Perspectives of Channelized and Unchannelized Headwater Streams in Central Ohio: A Case Study. *Journal of Water and Health*, **8**, 577-592. https://doi.org/10.2166/wh.2010.160

4

Effects of Co-Existing Ions on the Phosphorus Potassium Ratio of the Precipitate Formed in the Potassium Phosphate Crystallization Process

Hiroyuki Harada[1*], Yumi Katayama[1], Asmak Afriliana[2], Minori Inoue[1], Ryota Teranaka[1], Yoshiharu Mitoma[1]

[1]Faculty of Environmental Science, Prefectural University of Hiroshima, Hiroshima, Japan
[2]Faculty of Agricultural Technology, Univerity of Jember, Jember, Indonesia
Email: *ho-harada@pu-hiroshima.ac.jp

Abstract

Livestock wastewater is mainly treated with activated sludge, but ions such as phosphorus, potassium, ammonium, nitrate and sulfate remain in the effluent. In this study, the effects of residual ions on phosphorus recovery using the magnesium potassium phosphate crystallization method were investigated when magnesium was added to increase the pH. If co-existing ions affect the products, the phosphorus to potassium molar ratio (K/P ratio) of the precipitate will deviate from being equimolar. Artificial wastewater test solutions containing 5.6 - 20.3 mM ammonium, 25.6 mM potassium, 6.5 mM phosphorus, 0 - 7.35 mM nitrate, and 0 - 3.06 mM sulfate were used. First, the optimum operating pH and amount of magnesium added to give a high phosphorus removal rate and recovery rate were determined. The experimental setup was a 10 L aerated and stirred reactor, and a 5 L settling tank. The K/P ratio in precipitate was approximately 1 using the optimum conditions. Continuous 2 h treatment allowed a white precipitate containing about 30 g of needle-like crystals to be obtained. Next, the effects of varying the ammonium, nitrate, and sulfate ion concentrations in the artificial effluent were investigated. Ammonium and sulfate ion concentrations of 8 mM or more and 3 mM or more, respectively, caused the K/P ratio to decrease to about 0.7 and 0.5, respectively. Varying the nitrate concentration did not affect the K/P ratio, even at a nitrate concentration of 7.35 mM.

Keywords

Magnesium Potassium Phosphate, Ammonium, Sulfate, Nitrate, K/P Ratio, pH, Component, Formatting, Style, Styling

1. Introduction

Modern agricultural systems require phosphorus compounds to be applied in fertilizers, but the raw materials required (phosphorus ores) are likely to be exhausted at current rates of use. According to Kuroda *et al.*, The 80% of the phosphorus ore that is extracted is currently used in fertilizers [1]. Consumption is currently increasing at 3% each year. If consumption continues to increase at this rate, consumption in the 2060s will be about five times current consumption. This will result in economically minable phosphorus ore reserves being depleted. A realistic scenario is that 50% of currently minable ore reserves will still be available in the 2060s. Japan relies on imported phosphorus ore [1]. It is therefore desirable for techniques for recovering P from waste to be developed to decrease the risk of ore reserves being depleted and decrease the reliance of Japan on imported phosphate ore. Techniques for recovering phosphorus currently discharged into the environment are required to prevent environmental degradation. Livestock wastewater contains high concentrations of phosphorus, so it is easy to study phosphorus recovery methods using livestock wastewater. A traditional livestock wastewater treatment process is shown in **Figure 1**. First, solids, such as feces, in the wastewater are separated from the liquid. The liquid is then treated with activated sludge, which decomposes the organic substances and nitrifies the ammonium, turning some into nitrate. Even after these treatments the liquid will contain a high concentration of phosphorus.

The ammonia concentration is high at the start of the biological treatment process, so phosphorus can be recovered as magnesium ammonium phosphate (MAP; $MgNH_4PO_4$) using coexisting Mg. The crystallization of MAP is caused by a reaction among PO_4^{3-}, NH_4^+, and Mg^{2+} in aqueous solution. The reaction is shown in formula (1).

$$Mg^{2+} + NH_4^+ + PO_4^{3-} \rightarrow MgNH_4PO_4 \quad (1)$$

The MAP precipitation method is used for liquids containing P and ammonium. Mg^{2+} is added to generate MAP in weak alkaline solutions. The MAP method gives a high crystal formation rate, so crystallization occurs without seeding. This MAP method can be used at pH 8.5 to pH 9. The MAP method is of-

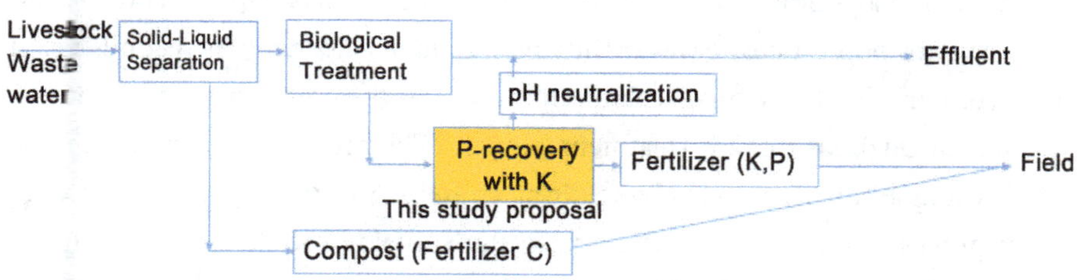

Figure 1. Schematic representation of a traditional livestock wastewater treatment process.

ten used for solutions containing relatively high P or ammonium concentrations, such as livestock wastewater and side stream wastewater produced during activated sludge treatments [2]-[7].

The ammonium concentration will decrease and MAP treatment will be difficult when a long treatment time is used and the nitrification efficiency is high. In this study, we investigated the recovery of phosphorus as magnesium potassium phosphate (MPP) through the reaction shown in formula (2) using a high potassium concentration.

$$Mg^{2+} + K^{+} + PO_4^{3-} \rightarrow MgKPO_4 \quad (2)$$

The MPP crystallization method has been used for wastewater with a high phosphorous concentration (30 mM or more, e.g., in urine) and a high potassium concentration [8]. Wastewater treated using activated sludge contains a low ammonium concentration, so it is difficult to use the MAP method to remove phosphorous from treated water. However, the relatively high K concentrations in wastewater treated using activated sludge can be appropriate for MPP treatment. The MPP treatment method has been used to remove relatively low phosphorus concentrations from wastewater treated using activated sludge in very few studies. The optimum pH for MPP treatment depends on the target phosphorous concentration because hydroxy ions are involved in the solubilities of the products. The optimum pH for treating wastewater containing 30 mM P in synthetic urine to give an equimolar precipitate of P, Mg, and K is pH 8 [8] [9], and the optimum pH for treating synthetic wastewater containing 7.6 mM P is pH 10 [10]. In the study described here, a P concentration of 6.5 mM was used to approximate the typical concentration in livestock wastewater, and the effects of the pH, Mg dosage, and coexisting substances on the formation of MPP were investigated using an experimental system that consisted of a reactor and a settling tank. The synthetic wastewater was designed to simulate effluent from piggery wastewater treated with activated sludge [11].

2. Methods

2.1. Apparatus and Standard Experimental Conditions

The experimental setup is shown in **Figure 2**. The system consisted of a 5 L raw water storage tank, a 10 L reaction tank, and a 5 L sedimentation tank. The synthetic livestock wastewater (containing potassium chloride, sodium dihydrogenphosphate, and ammonium chloride) in the raw water tank was fed into the reaction tank at a flow rate of 7.7 L/h. A dosage Mg/P as phosphate (PO_4-P) molar concentration ratio of 1 was maintained in the reactor tank by feeding a solution containing 50 mM magnesium chloride into the reaction tank at a rate of 1.5 L/h throughout the 2 h treatment process. The mixture was stirred and aerated using an air pump supplying air at a rate of 30 L/h throughout the treatment process. The basic synthetic wastewater contained K, PO_4-P, and NH_4^+ at concentrations of 25.6 mM, 6.5 mM, and 5.6 mM, respectively, and the pH was 7.5.

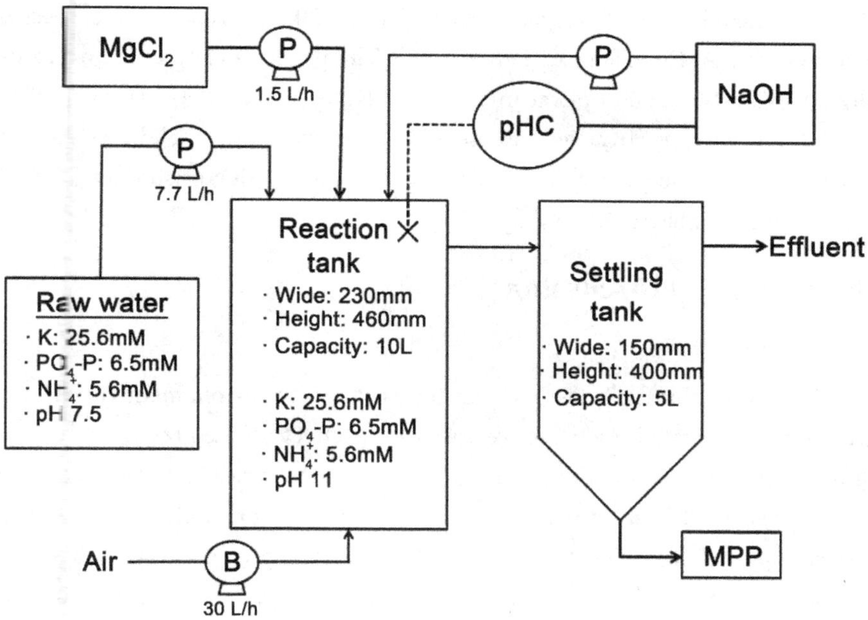

Figure 2. Experimental setup.

2.2. Effects of Varying the pH and Mg Dosage on the Formation of MPP

The pH of the solution in the reactor was varied between pH 8 and pH 12 using a pH controller. The dosage Mg/P ratio was changed from 1 to 2 by changing the rate at which magnesium chloride solution was added.

2.3. Effects of Coexisting Substances

Tests were performed using ammonium concentrations of 8.36 mM, 11.4 mM, and 22.3 mM at the standard conditions. Tests were also performed at nitrate concentrations of 0 mM, 4.02 mM, and 8.04 mM, and at sulfate concentrations of 1.02 mM, 2.04 mM, and 3.06 mM.

2.4. Precipitate

The white precipitate that formed in the reaction tank was moved to the settling tank in the overflowing solution and settled in the cone of the sedimentation tank. After a reaction, the mixture was allowed to stand for a specified time, then the reaction solution in the settling tank was passed through a filter. The white precipitate adhering to the filter paper was dried for 24 h in flowing air at 60°C. The filtrate was passed through a 0.45 μm membrane filter.

2.5. Analysis

The Magnesium concentrations in the samples were determined using an AA-6300 atomic absorption spectrometer (Shimadzu, Kyoto, Japan). The potassium and sodium concentrations were measured by flame photometry using the same atomic absorption spectrometer. The phosphate and ammonium concentrations were measured using the molybdenum blue and ascorbic acid method and using the indophenol method, respectively, and a V-530 ultraviolet-visible spectrophotometer (JASCO, Halifax, Canada) [12]. The precipitates were inspected using an ECLIPSE E-600 optical microscope (Nikon, Tokyo, Japan) and a JSM-6510 A scanning electron microscope (JEOL, Tokyo, Japan). The crystals were analyzed using energy dispersive X-ray spectrometry (EDS) using an EDX-LE instrument (Shimadzu, Kyoto, Japan).

3. Results and Discussion

3.1. Optimum pH

The MPP crystallization method can be used as a post-treatment for effluent treated with activated sludge, so we assessed the effects of the presence of nitrate, ammonium, and sulfate ions on the MPP production process. **Table 1** and **Figure 3** shows the Effects of pH control in component of precipitate, and Electron microscopy images of crystals produced. These components Mg, P, K of the crystals were almost equimolar at pH 11. Needle-like crystals peculiar to MAP and MPP were found at pH 9 or higher, but at pH 12 the components of the crystals stopped being equimolar because too much sodium hydroxide was present.

Needle-like crystals not seen at pH 8 became visible at pH 9, and the needle length increased and more needle-like crystals were present at higher pH values. The decreases in the P and K concentrations in the liquid increased as the pH increased from pH 8 and reached a maximum at pH 11, but then decreased when the pH was increased to pH 12. **Figure 4** shows the decreases in the concentrations of the components of the solution at different pH.

The amount by which the Mg concentration decreased did not decrease between pH 11 and pH 12, and the ammonium concentration remained almost constant.

Table 1. Effects of pH control in component of precipitate.

pH	Dosage Mg:P ratio	Mg/P (precipitate)	K/P (Preecipitate)
8	1:1	0.86	0.34
9	1:1	0.71	0.50
10	1:1	0.79	0.77
11	1:1	0.84	0.99
12	1:1	1.36	0.78

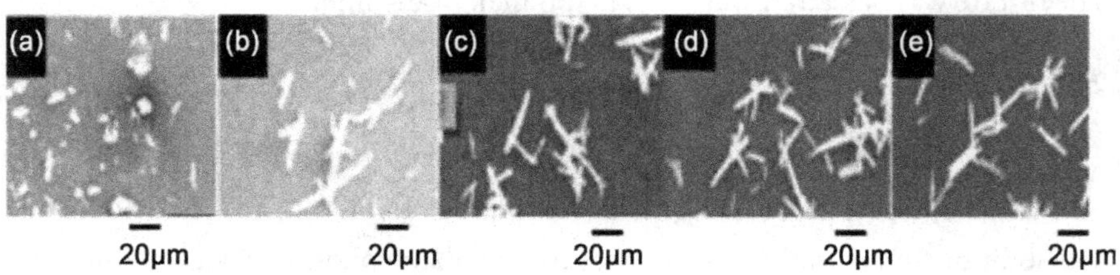

Figure 3. Electron microscopy images of crystals produced at different pH values. ((a) pH 8; (b) pH 9; (c) pH 10; (d) pH 11; (e) pH 12).

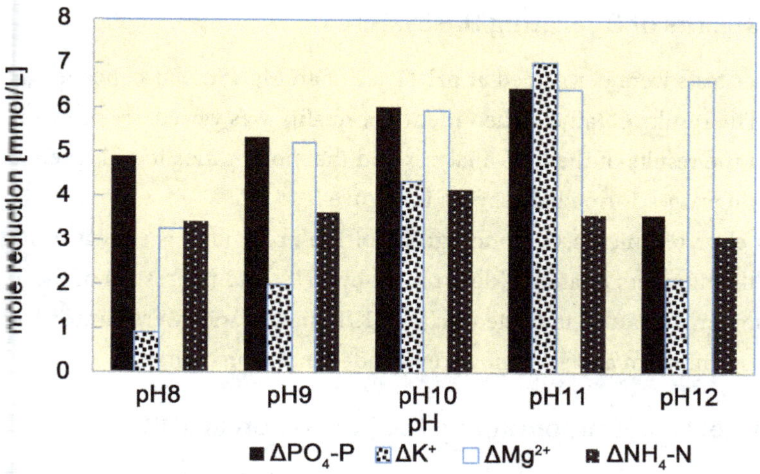

Figure 4. Decreases in the concentrations of the components of the solution at different pH.

Table 2. Phosphorus removal ratios and recovery ratios at different pH values.

pH	8	9	10	11	12
Removal rate (%)	74.8	82.0	92.5	98.8	55.1
Recovery rate (%)	93.2	96.1	97.7	95.5	99.7

At pH 11, the decreases in the phosphorous potassium, and magnesium Concentrations in the liquid were almost equimolar. The molar ratios of the precipitates, determined by EDS, were 0.84 and 0.95, and the ammonium and sodium contents were 4% and 8%, respectively. It therefore seems that the contents of the components of the precipitate reflected the decreases in the concentrations of the components of the solution. We will examine this using the removal rate and recovery rate using the equations below [13].

$$\text{Recovery rate of } PO_4\text{-P} = \frac{PO_4\text{-P in white precipitate}}{PO_4\text{-P decrement}} \tag{3}$$

$$\text{Recovery rate of } PO_4\text{-P} = \frac{PO_4\text{-P}_{\text{initial}} - PO_4\text{-P}_{\text{equilibrium}}}{PO_4\text{-P}_{\text{initial}}} \tag{4}$$

The results are shown in **Table 2**. The removal rate and recovery rate at pH 10 and pH 11 were 90% or higher. Finer crystals were produced at pH 11 than at pH 10, and this may have caused the recovery rates at pH 10 and pH 11 to be different. However, the difference was slight, so we conducted further tests at pH 11.

3.2. Optimum Dosage Magnesium

The effects of the magnesium dosage on the three components are shown in **Table 3**. The magnesium content of the precipitate was higher at a dosage Mg/P ratio of 1.3 than at lower ratios because magnesium hydroxide was precipitated. It appeared that the optimum magnesium concentration was the same as the phosphorus concentration.

3.3. Results of Operating the System

Further tests were performed at pH 11 using an Mg/P molar concentration ratio of 1. The results obtained when the water quality was varied are shown in **Table 4**, and the results of the EDS analyses and the molar ratios found when acid dissolution was performed are shown in **Figure 5**.

An electron microscopy photograph of the precipitate is shown in **Figure 6**. Both the numbers of atoms (determined by EDS) and the results of the acid dissolution analyses also indicated that the K/P molar ratios were almost 1 and that some ammonium and sodium were included in the precipitates.

3.4. Effects of Ammonium on the Formation of MPP

In our simple batch tests, ammonium concentrations up to about 5 mM did not affect the formation of MPP, and the K/P molar concentration ratios in the crystals were close to 1 [14]. The effects of varying the initial ammonium concentration on the formation of MPP were examined. The effects of varying the

Table 3. Effects of the magnesium dosage on the Mg:P molar concentration ratio in precipitate.

pH	Dosage Mg:P ratio	Mg/P (precipitate)	K/P (Preecipitate)
11	1.01:1	0.86	0.34
11	1.3:1	0.71	0.5
11	1.7:1	0.79	0.77
11	2.01:1	0.84	0.99

Table 4. Water quality parameters before and after the reactor process and in the settling tank and for the effluent.

Components	Concentration (mM) in the reaction tank		Concentration(mM) in the settling tank	Concentration(mM) in the effluent
	Before the treatment	After the treatment		
PO_4-P	6.48	0.21	0.16	0.14
K	25.6	18.8	19.0	19.6
Mg	0.07	0.44	0.23	0.15
NH_4^+-N	5.6	2.82	2.82	2.92

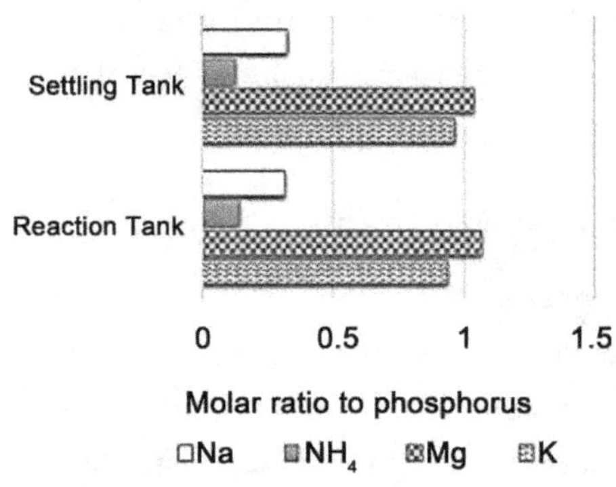

Figure 5. Results of the acid dissolution analyses of the components of the crystals.

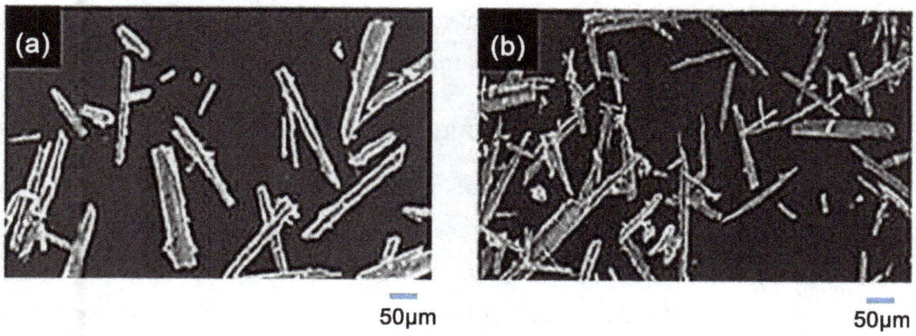

Figure 6. Electron micrographs of the crystals in (a) the reactor and (b) the settling tank.

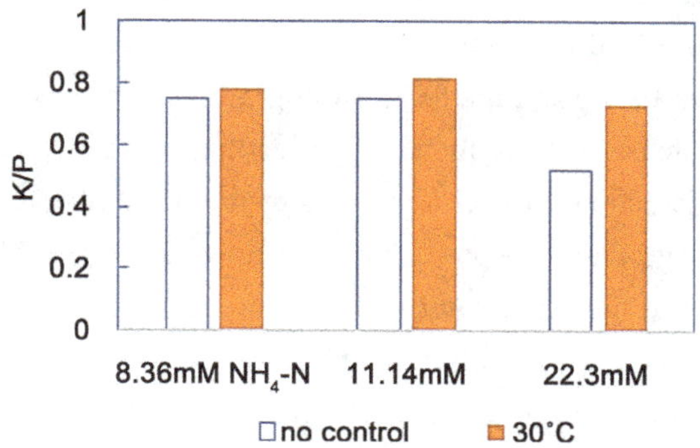

Figure 7. Effects of the initial ammonium concentration on the K/P ratio in precipitate dosage Mg/P = 1 at pH 11, 2 hours operation.

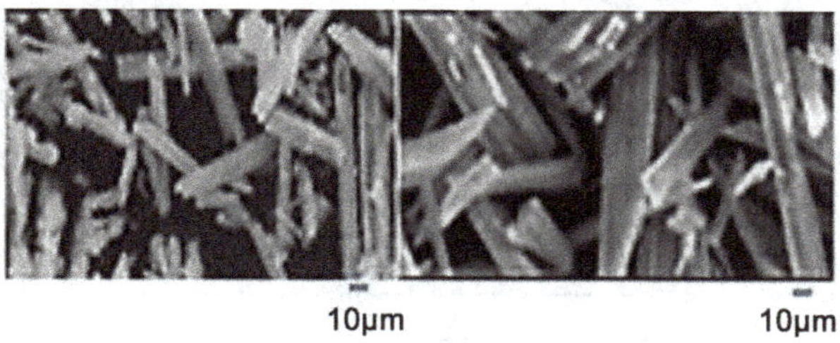

Room temperature 30˚C

Figure 8. Effects of temperature on the surfaces of the crystals formed at NH_4: 8.3 mM.

initial ammonium concentration on the K/P ratio are shown in **Figure 7**.

Electron microscopy photographs of the crystals produced at initial ammonium concentrations of between 0 mM and 22.4 mM are shown in **Figure 8**. In our simple batch tests, ammonium concentrations up to about 5 mM did not affect the formation of MPP, but higher concentrations caused the K/P molar concentration ratios of the crystals to be <1. Initial ammonium concentrations of 8.36 mM and 11.4 mM gave crystals with K/P ratios of 0.75 - 0.82, but the differences between the ratios found when the tests were performed at room temperature and 30˚C were small.

An initial ammonium concentration of 22.3 mM gave crystals with a K/P ratio of 0.73, but the ratio was only 0.52 when the test was performed at room temperature. It is therefore possible to decrease the effects of the presence of ammonium even at high ammonium concentrations by aerating and stirring the solution and controlling the temperature of the system. Even though the production of MPP was affected by the higher ammonium concentrations, needle-like crystals were found in the precipitate formed at each ammonium concentration that was tested.

3.5. Effects of Sulfate on the Formation of MPP

The effects of sulfate on the components of the solutions and precipitate and an electron micrograph of the precipitate are shown in **Figure 9** and **Figure 10.** We added nitrate to the wastewater with the compositions shown in **Figure 2** to investigate the effect of the presence of nitrate on the system. The tests were performed at nitrate concentrations of 0mM, 4.02mM and 8.06 mM. The decrease in the concentration of nitrate ion in the liquid is almost unchanged at 0.1 mM. The ratio of K to P in the crystal is also almost unchanged at 1: 1, and no difference in shape can be seen in the SEM photograph. From these points, it is considered that nitrate is no effect on MPP production.

The effects of sulfate on the components of the solutions and precipitate and an electron micrograph of the precipitate are shown in **Figure 11** and **Figure 12.** We added sulfate to the wastewater with the compositions shown in **Figure 2** to investigate the effect of the presence of sulfate on the system. The tests were performed at sulfate concentrations of 0 mM, 1.02 mM, 2.04 mM, and 3.06 mM. The white precipitate that was produced when the initial SO_4^{2-} concentration in the wastewater was 3.06 mM was analyzed by EDS and was found to contain 36.0% PO_4-P, 18.4% K, and 45.6% Mg. Little evidence of a sulfur component peak was found. Needle-like crystals were found at all of the sulfate ion concentrations that were tested. We therefore concluded that sulfate is not incorporated as a sulfur-containing compound in the crystals but inhibits MPP crystallization. No previous studies of the effects of SO_4^{2-} on MPP crystallization have been described,

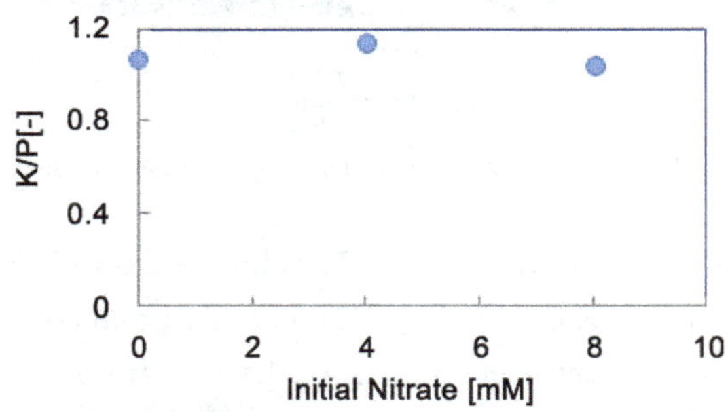

Figure 9. Effects of the initial nitrate concentration the K/P ratio.

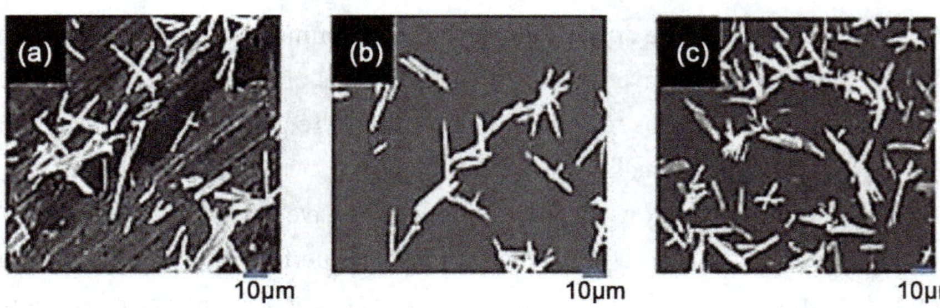

Figure 10. Effects of the nitrate concentration on the crystal on surfaces ((a) 0 mM, (b) 3.68 mM, (c) 7.35 mM).

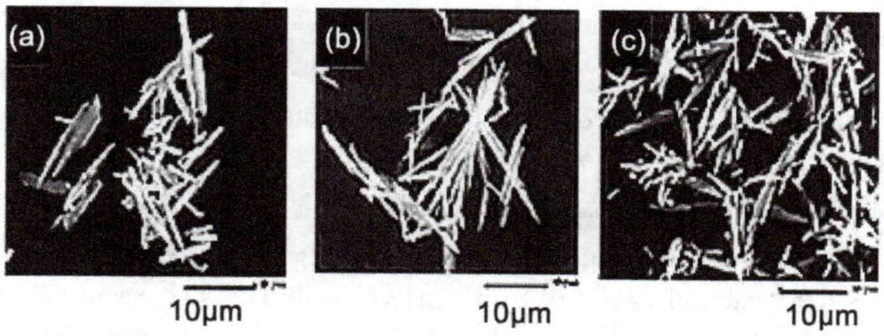

Figure 11. Effects of the initial sulfate concentration the K/P ratio.

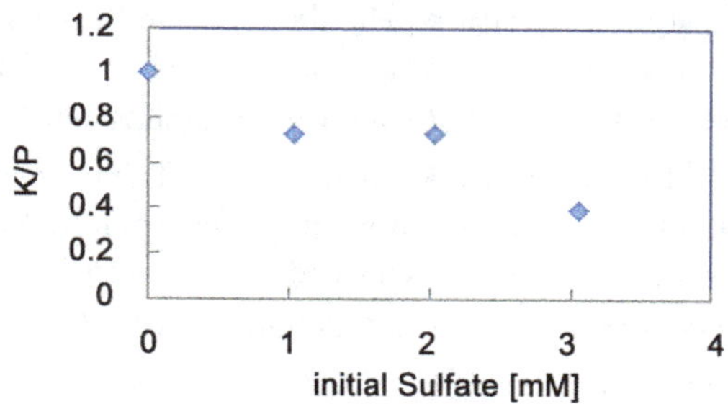

Figure 12. Effects of the sulfate concentration on the crystal on surfaces ((a) 1.02 mM, (b) 2.04 mM, (c) 3.06 mM).

but MAP production has been found to be affected by the presence of SO_4^{2-} at a concentration of 1300 mg/L (13.54 mM) [13]. It appears that the production of MPP is affected by the presence of SO_4^{2-} in the same way.

4. Conclusion

The aim of the study was to further improve livestock wastewater treatment and the recycling of phosphorous resources. The MPP crystallization process was applied to model wastewater containing 25.6 mM potassium, 6.5 mM phosphorus, and 5.6 mM ammonium at pH 11 using a Mg dosage equimolar with the P concentration. Crystals containing equimolar K/P ratios were obtained after 2 h of continuous treatment. Small amounts of ammonium and sodium were found in the crystals. Nitrate did not affect the K/P ratio, but ammonia and sulfate did affect the ratio, 1 mM sulfate in particular decreasing the K/P ratio to 0.7.

Acknowledgements

This work was supported by JSPS KAKENHI (grant no. 15K00605).

References

[1] Kuroda, A., Takiguchi, N., Kato, J. and Ohtake, H. (2005) Development of Technologies to Save Phosphorus Resources in Response to Phosphate Crisis. *Journal of Environmental Biotechnology*, **4**, 87-94. (in Japanese)

[2] Hirasawa, I., Tanaka, T. and Iwai, N. (1983) Studies on Phosphorus Removal by Contact Crystallization from Wastewater. *Japan Journal of Water Pollution Research*, **6**, 229-235. (in Japanese)

[3] Kaan, Y. and Zehra, S.-Z. (2009) Recovery of Ammonium Nitrogen from the Effluent of UASB Treating Poultry Manure Wastewater by MAP Precipitation as a Slow Release Fertilizer. *Journal of Hazardous Materials*, **166**, 260-269.

https://doi.org/10.1016/j.jhazmat.2008.11.025

[4] Sibel, U.-D. and Maazuza, O. (2009) Removal of Ammonium and Phosphate from the Supernatant of Anaerobically Digested Waste Activated Sludge by Chemical Precipitation. *Bioresource Technology*, **100**, 3236-3244. https://doi.org/10.1016/j.biortech.2009.02.030

[5] Song, Y-H., Qiu, G.-L., Yuan, P., Cui, X.-Y., Peng, J.-F., Zeng, P., *et al.* (2011) Nutrients Removal and Recovery from Anaerobically Digested Swine Wastewater by Struvite Crystallization without Chemical Additions. *Journal of Hazardous Materials*, **190**, 140-149. https://doi.org/10.1016/j.jhazmat.2011.03.015

[6] Suzuki, K., Tanaka, Y., Kuroda, K., Hanajima, D. and Fukumoto, Y. (2005) Recov ery of Phosphorus from Swine Wastewater through Crystallization. *Bioresource Technology*, **96**, 1544-1550. https://doi.org/10.1016/j.biortech.2004.12.017

[7] Suzuki, K., Tanaka, Y., Kuroda, K., Hanajima, D., Fukumoto, Y., Yasuda, T. and Waki, M. (2007) Removal and Recovery of Phosphorus from Swine Wastewater by Demonstration Crystallization Reactor and Struvite Accumulation Device. *Bioresource Technology*, **98**, 1573-1578. https://doi.org/10.1016/j.biortech.2006.06.008

[8] Xu, K., Li, J., Zheng, M., Zhang, C., Xie, T. and Wang, C. (2015) The Precipitation of Magnesium Potassium Phosphate Hexahydrate for P and K Recovery from Synthetic Urine. *Water Research*, **80**, 71-79. https://doi.org/10.1016/j.watres.2015.05.026

[9] Xu, K., Wang, K., Wang, X. and Qian, Y. (2012) Laboratory Experiments on Simultaneous Removal of K and P from Synthetic and Real Urine for Nutrient Recycle by Crystallization of Magnesium-Potassium-Phosphate-Hexahydrate in a Draft Tube and Baffle Reactor. *Chemosphere*, **88**, 219-223. https://doi.org/10.1016/j.chemosphere.2012.02.061

[10] Ymaguchi, S., Ohura, S., Harada, H., Akagi, K., Mitoma, Y., Kawakita, H. and Biswas, B.K. (2013) Simultaneous Crystallization of Phosphate and Potassium as Magnesium Potassium Phosphate Using Bubble Column Reactor with Draught Tube. *Journal of Environmental Chemical Engineering*, **1**, 1154-1158. https://doi.org/10.1016/j.jece.2013.08.032

[11] Tanaka, T. and Miyamaoto, T. (2013) Simultaneous Recovery of Phosphorus and Po tassium from Livestock by an Electrolytic Crystallization Method. *Journal of Japan Society of Civil Engineers Series G Environmental Research*, **69**, 1-9.(in Japanese)

[12] Eugene, W., Rice, R.B.B., Andrew, D.E. and Lenore, S.C. (2012) Standard Methods for Examination of Water and Wastewater. 21st Edition, American Public Health Association, Washington DC.

[13] Shih, Y.-J., Ruffel, M.A.R., Daniel, G.de L.M., Huang, Y.-H. and Lu, M.-C. (2017) Recovery of Phosphorus from Synthetic Wastewaters by Struvite Crystallization in a Fluidized-Bed Reactor: Effects of pH, Phosphate Concentration and Coexisting Ions. *Chemosphere*, **173**, 466-473. https://doi.org/10.1016/j.chemosphere.2017.01.088

[14] Akagi, K., Teranaka, R., Harada, H., Katayama, M.Y. and Biswas, B.K. (2016) Gen eration of Magnesium Potassium Phosphate from Synthetic Livestock Wastewater Containing High Levels Phosphorus and Potassium. *Journal of Water Pollution and Treatment*, **1**, 1-7.

5

The Impact of Water Price on the Financial Sustainability of the Palestinian Water Service Providers

Abdullah Murrar[1,2], Ibrahim Awad[3], Abdel Fattah Hasan[4], Eyad Yaqob[5], Ihab Barghothi[5], Ahmad Sadaqa[5], Subhi Samhan[6], Abdelrahman Tamimi[5,7]

[1]Faculty of Graduate Studies, Strategic Planning & Fundraising, Arab American University Jenin, Jenin, Palestine
[2]Tariff & Licensing Directorate, Water Sector Regulatory Council, Ramallah, Palestine
[3]Faculty of Economics and Business Administration, Al-Quds University of Jerusalem, Jerusalem, Palestine
[4]Water and Environmental Studies Institute, An-Najah National University, Nablus, Palestine
[5]Graduate Studies Teaching Committee, Arab American University Jenin, Jenin, Palestine
[6]Research and Development Directorate, Palestinian Water Authority, Ramallah, Palestine
[7]General Directorate, Palestinian Hydrology Group, Jerusalem, Palestine

Email: abdullah.murrar@gmail.com

Abstract

The impactof water sales price on the performance of water service providers is typically something of a mystery. High prices mean more revenue and profit; but it may lead to less bills collection and encourage the illegal connections. Yet, this argument has not been fully addressed in the Palestinian water sector; this research evaluates the effect of average water prices on the financial sustainability key indicators as collection efficiency, profit or loss percentage, non-revenue water, staff productivity, daily consumption, operating, and maintenance cost. The average price of cubic meter sold is segmented into low, medium, and high categories. Multivariate analysis shows that there are significant differences in profit or working ratio, daily consumption, and operating cost based on the different price categories. Further significant differences have been found in non-revenue water, collection efficiency, and water production based on low and high price categories. On the other hand, no significant difference has been found in staff productivity. The results show high price set by Palestinian water providers, leads to an increase in the bill collection rate and profit margin. However, negative relationship has been found between the price on one hand, and non-revenue water, average daily consumption, and water production on the other hand. The implication of these findings reveal that the Palestinian water providers should increase water prices gradually to cover operating and maintenance cost for better financial performance and sustainability.

Keywords

Collection Efficiency, Financial Sustainability, Palestinian Water Service Providers, Water Price, Working Ratio

1. Introduction

In the water sector, searching for a method to measure the performance, expanding services, and sustainability of water service providers has been one of the main concerns of regulatory bodies, non-profit organizations, and government entities around the world. As is the case in the electricity and communication sectors, water sector has received very special attention and interest; not only because water is a must for every human being, but also because of its nature of operation, management, and structure as a natural monopoly.

To measure the performance of water service providers, the researchers and water experts have developed performance indicators based on international standards. The International Water Association IWA, a hub of the water sector in the World has been facilitating the work of experts and professionals for the purpose of finding creative solutions for current water problems for the past 60 years. The performance indicators that are published by IWA have always provided effective tool to reflect the current situation, trend, and tracking it over time for water service providers.

In essence, the number and choice of the indicators are important since they give full picture about different performance areas of water service providers. The financial management, profitability, and bill collection efficiency have received considerable interest since they are core indicators. The profitability can be measured by the working ratio. This ratio is simply calculated to be the total amount of operating and maintenance expenses incurred by service provider divided by the total operating revenue generated during the year.

A surplus in the income statement of the water service provider without change this surplus into cash inflow would be difficult for water service provider to continue providing water services to the customers. Cash inflow is the lifeblood of the water providers. It's important because it later is converted into payment for things that make providing water services stay in business; such as administrative expenses, employees, rent, and other operating expenses. Therefore, the ability of the service provider to collect the water bills from its customers is an indicator for the management efficiency. The more the invoices payment promptly by the customers, the more the financial sustainability for water services providers. Naturally, pure positive cash flow is preferred.

In some countries, the efficiency of water bill collection may be compared with other performance indicators of the service providers to make sure that there is an effect on the overall performance. Some researchers find correlation between the customer's payment and non-revenue water. Low level of customers' payment causes increasing non-revenue water level; since there is little or no incentive to save the water when no intention to pay for it.

Given the fact that generating profits, collection efficiency, and non-revenue water reduction are key aspects of maintaining water service providers' financial sustainability; the water providers may achieve profitability by increasing price or water tariff, decrease cost and administrative expenses, or do both at the same time. The higher the prices and the lower the cost, the better chance to achieve financial sustainability. Therefore, financial sustainability is a matter of collection receivables, profitability, covering operating expense, ability to allocate for capital investment, minimal level of non-revenue water, and payment to water bulk supplier. All those financial performance areas have direct effect on water prices and vice versa. The purpose of this study hence is to measure whether significant differences in the financial performance based on the different price categories of water services; and which areas that always will be affected. The implication of the findings will support the decision makers for better financial performance and sustainability.

2. Palestinian Water Providers

According to the data bank of Palestinian Water Authority and Water Sector Regulatory Council [1], there are more than 280 water and wastewater service providers in the form of water and wastewater utilities, undertakings, authorities, water departments within municipalities, village, joint service council's, cooperative associations, or private sector. Since there is only one cooperative association in AbuDis, and one private sector provider in the newly established city of Rawabi; this study will focus on the three types in terms of institutional structure and ownership. Firstly: Regional Utilities, those are semi-independent and report to their board of directors. Joint Service Councils are reporting to ministry of local government directly; and Water Department within the Municipalities report to mayor of municipality, which at the end reports to ministry of local government. The Palestinian Water Law 2014, [2] calls for merging current water providers into regional utilities; changing structure and ownership from municipalities to be fully legal and financial independent utilities. The purpose of this merging is based on expectation that amalgamation will achieve more efficiency, high quality of water services, direct monitoring, expanding the services into new areas.

3. Palestinian Water Tariff

A tariff for water and wastewater services, is the tool to set the appropriate water price a user of these services is expected to pay. The consumer of water and wastewater services may be of either a low or high level of income; therefore, a different price blocks are set to achieve the user appropriate price [3]. In Palestine, the water tariff bylaw has determined that water price shall be based on incremental tariff blocks [4]. According to Article 4, the price shall be increased as there is an increase in the water consumption. Further, the wastewater tariff shall also be incrementally based on water consumption. This means that the more the water consumption, the higher the price of each cubic meter for water and also for disposal of the wastewater.

The tariff blocks in Palestine are four according to tariff bylaw 2013. The first block is up to 10 cubic meters; the second block is from 10 to 20 cubic meters. The third one is from 20 to 30 cubic meters for high consumption issue, and the last block is more than 30 cubic meters. The current implemented tariff in water service providers are always different; few of them have adapted to this structure. On the other hand, there is a fixed charge that shall be paid for each connection without consideration to consumption. Fees always are set to cover cost of connections, maintenance, and operations [4].

The tariff structure has many objectives as cost recovery, financial sustainability, efficient allocation of scarce resources, and income distribution [5]. The most carefully designed tariff cannot accomplish all of these objectives together; low level income may affect financial sustainability and cost recovery. Trading off and balancing between those objectives are optimal method in tariff set.

An important consideration often overlooked during developing pricing system is the efficiency of the operations. Customers will react favorably to good service and will be willing to pay for it with less interest in the price. An empirical study in Uganda showed that customer satisfaction and serveries quality *i.e.* not water price contribute significantly in the behavior of the customer to pay water invoice [6]. Conversely, poor service will evoke the public opposition to new or revised tariffs and payment water bills. A study shows there is no relationship found between customers' income, water price, and nonpayment of water invoices [7]. A measure of efficiency often used non-revenue water, which is the difference between the quantity of water supplied into the network and the quantity of water consumed whether metered or not. It is primarily the result and aggregate of leakages, illegal connections, metering inaccuracies, and unbilled consumption. Therefore, some water regulatory bodies, government entities accept specific level of non-revenue water during water prices approval.

4. Integrative Review

Many literatures have tackled water tariff structure and system. However, limited researches have concentrated average price and its effect on financial performance of water service providers. Some researchers find that high water price causes more revenue for water service providers, expanding water services into new areas, capital investment as pumping and networks, therefore, more in financial sustainability. Others may argue that low water prices encourage customers to pay their water bills since the amount is small compared to their income. It reduces the illegal connections and prevents social unfair use of water, because it becomes available at reasonable and affordable prices. It is imperative, therefore, to review related studies which to that end lead to exert the effect of water price on different performance areas.

Baietti, Kingdom and Ginneken, [8] studied the characteristics of well-performing public water utilities. The researchers conclude that many performance areas water provider can enhance to be rated as good performance such as tariff efficiency. The water utilities have not only cover operating and maintenance costs; but also a majority to generate a profit or surplus large enough to service their account payable and allocate given amount toward new investments for expansion of services. The tariff shall also be fair when compared with per capita income, so measuring ability of low level income to pay water bills.

Murrar, Tamimi & Samhan, [9] investigated the determinants of non-revenue water and financial viability for the Palestinian water service providers. The authors collected many parameters that affect non-revenue water and financial viability of the Palestinian water providers. Two multiple regressions have been conducted. One of the predictors for those regressions was the average price. The findings of this measurement show that, average price has significant impact on non-revenue water. Positive relationship between price and consumption from one side, and the financial viability from the other side. The low in price and less in quantity sold, results less in revenue generated by water service providers, which to that end leads to less in financial viability. In high non-revenue, low water prices, those conditions lead to insufficient amount of generated revenue, and therefore bad financial performance. The results of this research show high effect of price on non-revenue water and financial viability. The increasing price by one unit, results in decreasing non-revenue water by 0.346 units, other things being equal. On the other side, increasing price by one unit, results in more of financial viability by 0.821 units, other things being constant. This explains high price leads to generate more revenue, and then more in good financial performance and profitability.

Abdullah Murrar, [10] studied another performance area of Palestinian water service providers which is the collection efficiency of water bills, and motivational strategies that affect and encourage the Palestinian water customers to pay water invoices. Primary data has been collected from water experts and staff in the water service providers. The descriptive and inferential analysis have been conducted on the collected data. The findings of multiple regression showed that strategies of late payment penalties, early payment discount, and incremental tariff blocks are not significantly associated with this motivation. This means that if water service provider decides to decrease the water price, then the collection efficiency may not be increased. The Palestinian water customers will not pay their due invoices as a result of low price. This was in conformity with another study in China, where the price was not determinant in the customer's payment due invoices rather than the quality of services. The study showed that current price of irrigation water is too low and therefore it can't achieve sustainable use of water. The main reason is not farmer inability to pay, but unwillingness to pay due to poor services founded in the management of the water [11].

On the opposite direction, some water service providers add penalties on late payment of water customers. Or even increase significantly water tariff to subsidize uncollected invoices. The clue to this increase is revenue generating from issuing invoices will be increased. The invoice amount will become more than before and this may not encourage customers to pay their invoices. On the other hand, the government entities that supposed to approve high tariff rate may not agree with high tariff blocks due to reserve low income households. In Kenya, a study over water pricing and poor showed that high-income households and non-residential customers receive a disproportionate share of subsidies due to water tariff and that subsidy shall target poor household [12].

Abdullah Murrar *et al.*, [13] studied the efficiency and institutional performance of the Palestinian water service providers. The collected cross sectional data covers the period from year 2010 up to year 2015. The researchers found that the sales price of cubic meter plays major factor in revenue calculation. The data showed that on average Joint Services Councils "JSCs" deliver water services at price of 5.5 NIS, where water department in municipalities can charge people only by 3.2 NIS on average, however, the regional utilities set their services at 4.7 NIS. On the other hand, for the cost of water cubic meter, the same cost for both JSC and utilities *i.e.* 5.16 NIS, where, it cost less for the municipalities by 3.88 NIS. This means that when moving from dependent to autonomy structure of Palestinian water providers *i.e.* from municipalities to utilities and to JSCs; the water price raises, the gross profit margin increases and achieves more coverage of operating and maintenance cost.

Rubio, Villaverde & Gómez, [14] analyzed the ability of urban water tariffs in Spain to recover costs and to promote efficiency, sustainability, affordability and equity. The researchers found the amount of water bills forms small percentage of family income. However, many families become unable to pay the due water invoices. Mathur & Vijay, [5] found that customers give little or no attention towards conservation of water since its inexpensive and therefore this encourages people to waste. This means that the water price plays an effective tool for customer behavior. The higher the water sales price are, the more negative behavior may appear through illegal connection and/or no payment the water invoices.

5. Research Methodology

A review of related studies clearly elucidated that there are many procedures and strategies that can be adapted by water service providers to enhance their financial performance. This research will test by conducting Multivariate analysis of variance (MANOVA) the relationship and significant impact of price as predictor over many dependent parameters as staff productivity, daily consumption, energy cost, service provider size, service provider structure, non-revenue water, collection efficiency, and profit or loss percentage of Palestinian water service providers. When one predictor and many respond variables, the MANOVA technique is used to test whether there is significant difference between groups based on the independent variable [15].

The research mainly depends on secondary data that has been collected from published performance indicator reports of Palestinian water service providers. The performance reports used to be published by Palestinian Water Authority (PWA); but currently are published by Water Sector Regulatory Council (WSRC); with full support as financial and technical advisors team by Deutsche Gesellschaftfuer Internationale Zusammenarbeit (GIZ) Water Program. According to the WSRC, the published data in 2016 report covers more than 70% of the total Palestinian population. Expressed in other terms, the sample size in this research will include all water service providers that deliver water services to more than 70% of the Palestinian population. The sample size of this research contains three Palestinian water regional utilities, 5 joint service councils, and 55 water departments in the municipalities; this comes to 63 Palestinian water service providers. However, to enhance the representative sample, a cross sectional data will be considered; where, this paper includes all data in performance reports from year 2010 and up to year 2015 for all service providers.

Those observations will be analyzed and tested using Statistical Package for Social Science (SPSS). Both descriptive and inferential analyses will be carried out. The purpose of this inferential test is to know whether significant differences appeared in those performance dimensions based on the average price of water providers.

In this paper, the water prices are classified into three broad categories based on the current water average prices. The average price of cubic meter is calculated by dividing the total net sold quantity during the year by water service provider over the number of cubic meters invoiced. The low level category includes range from lowest price up to 33% of the highest average price. However, for the high category, it starts from 66% of the highest price, therefore, the medium category is in between low and high. **Table 1** summarizes the range and the number of water service providers in each category.

The performance areas that are affected by average water price are selected from those observations to achieve financial sustainability of service providers. Four elements have been considered: firstly, profit or loss generated by service provider. For this factor, operating and maintenance cost is considered. On the

Table 1. Price categories.

Price Category	Price Range NIS	Average Price NIS	No of Water Providers
Low	0.81 to 2.75	1.66	57
Medium	2.76 to 5.50	4.19	50
High	5.51 to 8.31	6.14	36

other hand, gross profit margin and working ratio have been included to support this factor. Generally, less cost and high price, produce more in gross and net profit margin. Secondly, consumption, the average consumption may be indicated as the ability of Palestinian service provider whether municipality, regional utility, or council to deliver a good quality of services and continuous supply. The consumption pattern always affected directly by the price and tariff structure.

Thirdly, the collection efficiency dimension. This aspect is correlated with the profit generating since, this profit shall be collected and changed into cash inflow to enable service provider to pay due invoices and operating expenses. The lower the collection efficiency, the less liquidity the water utility may experience. For this reason, collection efficiency has been reflected as a key performance indicator. The last dimension in this performance is service providers' efficiency in non-revenue water reduction and employees' productivity. The less the non-revenue water percentage, the more the management efficiency, and hence the financial sustainability.

To enhance the analysis especially the degree of relationship between variables, a correlation matrix is considered in this research since it is one of the most common and useful statistics. The correlation measures the strength and direction of linear relationships between price and other performance indicators as profit, working ratio, non-revenue water, collection efficiency and so forth. By extension, the correlation evaluates whether there is statistical evidence for a linear relationship among those variables in the population [16].

6. Data Analysis and Discussion

The cross sectional data is analyzed and tested using Statistical Package for Social Science (SPSS). Both descriptive and inferential analyses have been carried out. **Appendix Table A2** summarizes the collected data from performance reports of Palestinian water service providers. The table shows 143 observations for near to 65 water service providers. The data covers the period from year 2010 and up to year 2015. However, **Appendix Table A3** proposes multiple and significant comparisons between those prices. The approximate multivariate for Wilk's Lambda analysis as in **Appendix Table A2** shows that overall model is significant where $p = 0.000$ [17]. Therefore, there is a statistically significant difference in performance of water service providers based on the price categories, where $F = 20.620$, $p < 0.0005$ and Wilk's $\Lambda = 0.151$.

6.1. Profitability Analysis

The major performance area for water providers is the bottom line in their income statement. It is expressed by the working ratio; it equals total amount of operation and maintenance expenses incurred by water provider over operating revenue generated during the year. The purpose of this ratio hence is to measure the ability of water provider to cover operating and maintenance cost from revenues, and whether the remaining amount will cover capital allocation and investment. The default amount of this ratio is 1, wherein total revenue equals total operating and maintenance expenses. Less than 1 means revenue covers operating expenses with considerable margin. Back to **Appendix Table A1**, it shows the higher the water price of cubic meter, the less the working ratio will be. This can be explained as water providers can generate more operating profit compared with water providers that set their price at low and medium categories. Again, the operating cost per cubic meter for all water providers is considered in this calculation.

Appendix Table A3 shows that there is significant difference in working ratio between low and medium price categories, low and high, and therefore between medium and high. To gofarther in the analysis, the operating and maintenance cost is considered. **Appendix Table A1** sketches positive relationship between price category and cost per cubic meter. The higher the cost of cubic meter, the higher the average price. The results reveal that there is significant difference in operating and maintenance cost based on the three categories of water prices.

As a general rule, the working ratio and gross profit are affected by revenue and cost. The more in billed revenue and less in recognized cost, the more in profit achieved by water service providers. The sales price of cubic meter plays major factor in revenue calculation, therefore, it's necessary to set average price and average cost per unit sold, side by side with net profit and average consumption per capita per day.

Table 2 shows high loss rate incurred by water providers that set their prices in low category. The high category covers operating and maintenance cost without inclusion of deprecation. The correlation table proposes negative and significant relationship between working ratio and average water price from one side; and positive relationship with operating and maintenance cost from the other side. This implies that Palestinian water providers always set their prices based on the cost, but without full knowledge on the cost calculation. Those results are in conformity with previous study that showed fitted lines of average cost and average price. The researcher concluded that Palestinian water service providers, especially large providers always set water tariff based on calculated cost [18].

6.2. Average Consumption Analysis

The core service of water providers is to deliver good quality water to the customers. If there is no water to deliver, then, the customer will not receive the

Table 2. Price cost relationship.

Price Category	Average Price NIS	Average Cost NIS	Net Profit or Loss	Consumption l/c/d	Bills Collection
Low	1.66	2.67	−70.0%	116	52%
Medium	4.19	4.76	−10.0%	68	68%
High	6.14	5.84	5.00%	72	72%

required quantity and therefore the average daily consumption per capita per person will be at the low level. In Palestine, the customers in Qalqilia and Jericho service providers have an average consumption of 170 and 271 liters per person per day respectively. Where in Dahiriya and Yatta municipalities, the average consumption quantities reaches 31 liters per person per day. Leaving other things constant, the main reason for this variation between high and low consumption is water availability. In Qalqilia and Jericho, wherein large consumption quantities, the water providers are delivering in continuous supply mode. However, high intermittent supply is appearing in low consumption areas, *i.e.* the customers in Yatta and Dahiriya may wait for more than one month to receive their share of water especially in the summer season.

Appendix Table A1 shows high daily consumption in the areas of low water prices *i.e.* Qalqilia, Jericho and Zeita. However, when the average prices move into high category, the consumption decreases by 50%. The major reason for this decreasing in addition to high price is the water shortage in that areas. The multiple comparison table displays $p < 0.05$; this implies that there is significant difference in Palestinian customer consumption based on the three price categories.

6.3. Non-Revenue Water Analysis

Appendix Table A1 shows non-revenue water percentage is low in high price-segment, which means that it is high percentage in low price category. The non-revenue water is near to 36% in low price category, 31% in medium, where, its 28% in high category. From statistic point of view, there is significant difference in non-revenue water based on low and high price category; however, no significant difference is found between the other categories. In Palestinian water sector, the water providers that deliver water in low price have less interest to decrease the non-revenue water. Mathur & Vijay, [5] found that customers give little or no attention towards conservation of water since its inexpensive and available. However, when the cost of cubic meter is relatively high, no water available, the intension is to save the quantity of water and high cost, therefore the non-revenue water percentage is less.

The correlation table displays negative relationship between price from one side and non-revenue water from the other side. The more the daily consumption, the less the non-revenue water is. In Palestine, some water providers have non-revenue water projects, especially from international donors, so changing unmetered to be metered; this decreases the non-revenue water percentage. In North West Jenin Council, this Palestinian water service provider has non-revenue water 39%, number of connections is 6000, network length is about 505 km, capital expenditures by USD 500,000, from its own tariff has been disbursed to decrease the non-revenue water. Many activities have been followed such as nigh flow, changing some network pipes, meters, and so forth. Within two years, the non-revenue water becomes less than 20%.

From financial point of view, high price of cubic meter means covering cost and possibly allocating part of surplus for development projects including non-revenue water reduction projects. In this setting, the low in price and less in quantity sold, results less in revenue generated by water utility, which to that end leads to less in financial performance. A study shows high impact of price and consumption on non-revenue water. The increasing price by one unit, results in decreasing the non-revenue water by 0.346 [9].

6.4. Collection Efficiency Analysis

Showing surplus in the income statement of water service provider without change surplus into cash inflow; would be difficult for water service provider to continue providing water services to customers. Cash inflow is the lifeblood of the water providers. Its important because it later becomes payments for things that make providing water services run; such as administrative expenses, employees, rent, and other operating expenses. Therefore, the ability of water provider to collect water bills from its customers is an indicator for management efficiency. The more the invoice payment promptly by the customers, the more the financial sustainability for water services providers. Naturally, positive cash flow is preferred. Positive cash flow means water service provider is running smoothly. High positive cash flow is even better and will allow to make new investments; *i.e.* expand the water services into new areas, expand the water network, purchasing pumps and others. The negative cash flow refers to more money paying out than being coming from customers. That is why water bills collection is core indicator for sustainability and services continuity.

This research demonstrates that low price category of the Palestinian water providers may collect only 52% from their annual water bills; the medium category providers collect 68%, where high price category providers collect 72%. For the low price category, the water providers may collect only half of its annual revenue. The other part *i.e.* 48% of annual water sales will always be accumulated into next year as account receivables. The **Table 2** shows that loss in the price category reaches up to 70%. This implies that half of this revenue always not recovered and recognized due to collection problem.

In the Palestinian water sector, some of the strategies are always implemented in high price category of service providers. Those strategies lead to an increase in the collection efficiency such as installing prepaid meters instead of postpaid meters, implementation of an advanced technology such as mobile software, quality of water services provided, quality of other services provided, and customers' satisfaction. All those predictors motivate customers to pay their water bills (Murrar, 2017). **Appendix Table A4** shows positive and significant relationship between the collection efficiency and the average water sales price. The more water price for cubic meter, the more the collection efficiency.

6.5. Water Production Analysis

In Palestine, the water supply for service providers comes from generally two sources: water production and bulk purchases. It has been noted that water providers that depend on production rather than purchases such as Qalqilia and Zeita incurred less cost, charge low price to customers, and have more water available therefore, more daily consumption by consumers. **Appendix Table A4** shows highest and inverse relationship between the average price and dependency of water provider on production rather than purchasing. The more the dependency on production, the less the price of cubic meter charge. In Palestine, the bulk purchase price of cubic meter is near to 2.86 NIS; however, on average the cost of producing one cubic meter from own wells of water providers doesn't exceed 2.0 NIS. In Balkan countries, a study shows negative relationship between water production and the cost; *i.e.* the more the production and consumption, the less the cost per cubic meter [19].

The findings of this paper demonstrate that, the water providers which produce rather than purchase have incurred more non-revenue water and less operating and maintenance cost. This means that, there will be considerable quantity of water losses during the production and transmission process. Kingdom, Liemberger & Marin, [20] estimated from limited set of projects in developing countries, the unit cost of reducing physical leakage range from $215 to $550. Therefore, those experts who suggest the establishment of water utility should conduct cost benefit analysis before moving on with the decision. In Palestine, the water providers especially those produce rather than purchase such as Jericho and Qalqilia consider the cost and the effect of fixing physical leakage.

Low cost of water producers not necessary means more profit. The results presented in **Appendix Table A4** prove insignificant relationship between water providers that have private wells and the working ratio. The reason for that is simply that the water providers set their sales price of cubic meter at a low level without covering operation and maintenance cost. Another weak performance for this water provider's category is the deficiency of bill collection. The results show that water providers who have private wells are inefficient in bills collection from the customers comparing with the water providers that don't have own wells, and purchase from other resources. One reason for this variation is that the water providers shall pay water invoices to the bulk supplier "West Bank Water Department". The more the pressuring from the bulk supplier on water provider to pay or schedule the due invoices, the high the collections rate of water provider from the end customers.

7. Conclusions & Policy Implications

Studies investigated water tariff and its relation to different financial performance and social equality are abundant. Their importance lies in the fact that water prices play a major role in the financial sustainability and achievement in technical areas. Some authors find that high water prices cause more revenue for water service providers. This condition paves the way toward expanding water services into new areas, capital investment as renewal capital assets, therefore, higher efficiency and improved financial sustainability. However, others may argue that low water prices encourage the customers to pay their due water bills, decrease non-revenue water, illegal connections, and allowing people to consume the required quantity of water; since it becomes available at reasonable and affordable prices. The main purpose of this study hence is to evaluate the effect of independent parameter *i.e.* water sales price on different financial performance aspects. In addition, this study aims to come up with practical implication for decision makers whether encouraging increasing water price or keeping it at low level.

The result of this study indicates higher working ratio or profit can be generated for the water providers that set high price for each cubic meter sold. This means that there will be significant difference in profit achieved based on the low, medium, and high price categories. On the other hand, positive relationship between price categories and cost per cubic meter is found. The higher the cost of cubic meter, the higher the water average price. The implication for this argument is that the Palestinian water providers consider the cost in pricing strategy, but with no full coverage methodology. The descriptive statistic table shows high daily consumption in the areas of low water prices such as Qalqilia, Jericho, and Zeita. However, when the average prices move into the high category, the consumption is decreased by 50%. The major reason for this decreasing in addition to high price is the water shortage in those areas.

In Palestinian water sector, non-revenue water is an indicator for the efficiency of water provider. The providers that deliver water in low price have less interest to decrease the non-revenue water. However, when the cost of cubic meter is relatively high with water shortages, the intension becomes focused on saving the quantity of water and to decrease the high cost, therefore non-revenue water percentage is lowered. The non-revenue water is near to 36% in low price category, 31% in medium, where, its 28% in high category.

Other focus area on this paper hence, is the collection of water bills from customers. The more the invoice payment promptly by the customers, the more the efficiency and financial sustainability for water services providers. The results of those investigations show that low price category of the Palestinian water providers can collect only 52% from their annual water bills; the medium providers collect 68%, where, large providers collect 72%. This means the more the price per cubic meter, the more the collection percentage.

Appropriate costing and pricing mechanisms are important to enable of all water providers toward better financial performance and financial sustainability. Increasing water price gradually will enhance revenue stream for Palestinian water service providers, achieve surplus, decrease non-revenue water, and increase collection rate or percentage of water bills from customers.

Acknowledgements

This research paper has been made possible through the support from MEDRC

Water Research at the Sultanate of Oman. We would like to express our deepest appreciation to MEDRC for publishing this paper. The deepest gratitude also to Palestinian Water Authority (PWA) for valuable, effective supervision, coordination of the scientific researches, and directing them toward finding creative solutions for current water problems.

References

[1] Water Sector Regulatory Council (2014) Water Services Providers Performance Report 2014. Water Sector Regulatory Council. http://www.wsrc.ps/

[2] State of Palestine (2014) Decree No. 14 for Year 2014 Relating to the Water Law. Palestinian Water Authority.

[3] Laredo, D. (1991) Principles of Tariff Design for Water and Wastewater Services. Bureau for Science and Technology, US Agency for International Development under WASH Activity No. 380 and Task No. 141. Contract No. DPE-5973-Z00-808 1-00, Project No. 836-1249. Washington DC.

[4] State of Palestine (2013) Water Tariff Bylaw 2013. Council of Minister. Cabinet Secretaries. Setting Approval Number 37.

[5] Singh, U. and Mitta, A.K. (2005) Urban Water Tariff Structure and Cost Recovery Opportunities in India. *Water Science & Technology*, **52**, 43-51.

[6] Kayaga, F. and Sansom (2004) Bill Payment Behavior in Urban Water Services: Empirical Data from Uganda. *Journal of Water Supply: Research and Technology-Aqua*, **53**, 339-349.

[7] Vásquez and William (2015) Nonpayment of Water Bills in Guatemala: Dissatisfaction or Inability to Pay? *Water Resources Research*, **51**, 8806-8816. https://doi.org/10.1002/2014WR016610

[8] Baietti, K. and Ginneken (2006) Characteristics of Well-Performing Public Water Utilities. Water Supply and Sanitation Sector Board of the Infrastructure Network of the World Bank Group. Water Supply & Sanitation Working Notes. Note No. 9.

[9] Murrar, A., Tamim, A. and Samhan, S. (2017) The Determinants of Non-Revenue Water & Financial Viability for the Palestinian Water Service Providers. *Journal of Water Resources and Ocean Science*, **6**, 35-45. https://doi.org/10.11648/j.wros.20170602.12

[10] Murrar, A. (2017) The Water Invoices and Customer Payment Motivational Strategies: An Empirical Study on Palestinian Water Service Providers. *EPRA International Journal of Economic and Business Review*, **5**.

[11] Tang, N. and Liu (2013) The Willingness to Pay for Irrigation Water: A Case Study in Northwest China. *Global NEST Journal*, **15**, 76-84.

[12] Fuente, D., GakiiGatua, J., Ikiara, M., Kabubo-Mariara, J., Mwaura, M. and Whittington, D. (2016) Water and Sanitation Service Delivery, Pricing, and the Poor: An Empirical Estimate of Subsidy Incidence in Nairobi, Kenya. *Water Resources Research*, **52**.

[13] Abdullah, M., *et al.* (2017) The Efficiency and Institutional Performance of the Palestinian Water Service Providers. *American Journal of Environmental and Resource Economics*, **2**, 162-174.

[14] Rubio, M., Villaverde, A. and Gómez, F. (2015) Urban Water Tariffs in Spain: What Needs to Be Done? *Water Journal*, **7**, 1456-1479. https://doi.org/10.3390/w7041456

[15] Grice, J. and Iwasaki, M. (2007) A Truly Multivariate Approach to Manova. *Applied*

Multivariate Research, **12**, 199-226. https://doi.org/10.22329/amr.v12i3.660

[16] Veber Costa (2016) Correlation and Regression. In: Naghettini, M., Eds., *Fundamentals of Statistical Hydrology*, Springer, Cham.

[17] Haase, R. and Ellis, M. (1987) Multivariate Analysis of Variance. *Journal of Counseling Psychology*, **34**, 404-413. https://doi.org/10.1037/0022-0167.34.4.404

[18] Murrar, A. (2017) The Relationship between Size and Performance of Palestinian Water Service Providers. *Journal of Water Resource and Protection*, **9**, 536-552. https://doi.org/10.4236/jwarp.2017.95035

[19] Mathur, P. and Vijay, A. (2013) Non-Revenue Water Reduction—A Tool for Achiving 24x7 Water Supply. *IOSR Journal of Mechanical and Civil Engineering*, **7**, 25-28.

[20] Kingdom, B., Liemberger, R. and Marin, P. (2006) The Challenge of Reducing Non-Revenue Water (NRW) in Developing Countries How the Private Sector Can Help: A Look at Performance-Based Service Contracting. THE World Bank Group, Washington DC. PPIAF Water Supply and Sanitation Sector Board Discussion Paper Serial Paper No. 8.

Appendix

Table A1. Descriptive statistics.

	Price Category	Mean	Std. Deviation	N
Working Ratio	Low	1.5953	1.39800	57
	Medium	1.0226	0.21983	50
	High	0.8775	0.14948	36
	Total	1.2143	0.94493	143
Non-Revenue Water	Low	35.4163	10.20977	57
	Medium	30.9704	12.12426	50
	High	28.6394	11.19763	36
	Total	32.1557	11.42957	143
Staff Productivity	Low	4.9379	2.58508	57
	Medium	4.6030	2.57085	50
	High	4.9386	2.75610	36
	Total	4.8210	2.61038	143
Collection Efficiency	Low	52.0995	23.85310	57
	Medium	68.5330	28.71634	50
	High	72.1822	18.24631	36
	Total	62.9013	25.90054	143
O&M Cost	Low	2.6767	2.55240	57
	Medium	4.7590	1.63699	50
	High	5.8425	1.18458	36
	Total	4.2017	2.35907	143
Consumption	Low	116.2007	58.40227	57
	Medium	83.7160	42.97105	50
	High	64.8086	22.64441	36
	Total	91.9045	50.55069	143
Production Percent	Low	84.7728	33.11925	57
	Medium	30.4644	40.54653	50
	High	26.2272	37.20015	36
	Total	51.0451	45.85724	143
Location	Low	94.9924	49.36105	33
	Medium	87.0861	62.07258	59
	High	95.4808	34.37540	51
	Total	91.9045	50.55069	143

Table A2. Multivariate Tests[a].

Effect		Value	F	Hypothesis df	Error df	Sig.
Intercept	Pillai's Trace	0.987	1027.019[b]	10.000	131.000	0.000
	Wilks' Lambda	0.013	1027.019[b]	10.000	131.000	0.000
	Hotelling's Trace	78.398	1027.019[b]	10.000	131.000	0.000
	Roy's Largest Root	78.398	1027.019[b]	10.000	131.000	0.000
Price Category	Pillai's Trace	1.009	13.447	20.000	264.000	0.000
	Wilks' Lambda	0.151	20.620[b]	20.000	262.000	0.000
	Hotelling's Trace	4.564	29.669	20.000	260.000	0.000
	Roy's Largest Root	4.319	57.007[c]	10.000	132.000	0.000

[a]Design: Intercept + Price Category. [b]Exact statistic. [c]The statistic is an upper bound on F that yields a lower bound on the significance level.

Table A3. Multiple comparison.

Dependent Variable	(I) Price	(J) Price	Mean Difference (I-J)	Std. Error	Sig.	95% Confidence Interval	
						Lower Bound	Upper Bound
Working Ratio	L	M	0.5727	0.18776	0.010	0.1113	1.0340
		H	0.7178	0.18684	0.001	0.2584	1.1771
	M	L	−0.5727	0.18776	0.010	−L340	−0.1113
		H	0.1451	0.03984	0.001	0.0480	0.2422
	H	L	−0.7178	0.18684	0.001	−1.1771	−0.2584
		M	−0.1451	0.03984	0.001	−0.2422	−0.0480
Non Revenue Water	L	M	4.4459	2.18374	0.128	−0.8605	9.7523
		H	6.7769	2.30472	0.013	1.1379	12.4158
	M	L	−4.4459	2.18374	0.128	−9.7523	0.8605
		H	2.3310	2.53435	0.738	−3.8516	8.5135
	H	L	−6.7769	2.30472	0.013	−12.4158	−1.1379
		M	−2.3310	2.53435	0.738	−8.5135	3.8516
Staff Productivity	L	M	0.3349	0.49942	0.878	−0.8772	1.5470
		H	−0.0007	0.57292	100	−1.4017	1.4003
	M	L	−0.3349	0.49942	0.878	−1.5470	0.8772
		H	−0.3356	0.58582	0.920	−1.7675	L963
	H	L	0.0007	0.57292	1.00	−1.4003	1.4017
		M	0.3356	0.58582	0.920	−L963	1.7675
Collection Efficiency	L	M	−16.4335	5.14534	0.006	−28.9381	−3.9289
		H	−20.0827	4.38519	0.000	−30.7575	−9.4080
	M	L	16.4335	5.14534	0.006	3.9289	28.9381
		H	−3.6492	5.07352	0.854	−16.0131	8.7147
	H	L	20.0827	4.38519	0.000	9.4080	30.7575
		M	3.6492	5.07352	0.854	−8.7147	16.0131
Operating & Maintenance	L	M	−2.0823	0.40974	0.000	−3.0779	−L867
		H	−3.1658	0.39150	0.000	−4.1194	−2.2122
	M	L	2.0823	0.40974	0.000	L867	3.0779
		H	−L835	0.30426	0.002	−1.8248	−0.3422
	H	L	3.1658	0.39150	0.000	2.2122	4.1194
		M	L835	0.30426	0.002	0.3422	1.8248
Consumption	L	M	32.4847	9.83714	0.004	8.6044	56.3650
		H	51.3921	8.60713	0.000	30.3940	72.3902
	M	L	−32.4847	9.83714	0.004	−56.3650	−8.6044
		H	18.9074	7.15359	0.029	1.4514	36.3634
	H	L	−51.3921	8.60713	0.000	−72.3902	−30.3940
		M	−18.9074	7.15359	0.029	−36.3634	−1.4514

Continued

Production Percent	L	M	54.3084	7.21970	0.000	36.7596	71.8572
		H	58.5456	7.59499	0.000	39.9542	77.1369
	M	L	−54.3084	7.21970	0.000	−71.8572	−36.7596
		H	4.2372	8.44516	0.944	−16.3633	24.8377
	H	L	−58.5456	7.59499	0.000	−77.1369	−39.9542
		M	−4.2372	8.44516	0.944	−24.8377	16.3633
Water Provider Size	L	M	0.013	0.1466	000	−0.343	0.368
		H	−0.308	0.1595	0.161	−0.698	0.081
	M	L	−16.4335	5.14534	0.006	−28.9381	0.343
		H	−20.0827	4.38519	0.000	−30.7575	0.066
	H	L	16.4335	5.14534	0.006	3.9289	0.698
		M	−3.6492	5.07352	0.854	−16.0131	0.708
Water Provider Structure	L	M	−0.175	0.1034	0.261	−0.428	0.079
		H	−0.839	0.1491	0.000	−1.210	−0.468
	M	L	0.175	0.1034	0.261	−0.079	0.428
		H	−0.664	0.1719	0.001	−L86	−0.243
	H	L	0.839	0.1491	0.000	0.468	1.210
		M	0.664	0.1719	0.001	0.243	L86
Water Provider Location	L	M	0.907	0.1506	0.000	0.541	1.273
		H	0.917	0.1625	0.000	0.519	1.315
	M	L	−0.907	0.1506	0.000	−1.273	−0.541
		H	0.010	0.1800	000	−0.429	0.449
	H	L	−0.917	0.1625	0.000	−1.315	−0.519
		M	−0.010	0.1800	0.000	−0.449	0.429

Based on observed means. The error term is Mean Square (Error) = 0.604.

Table A4. Correlation matrix.

		WRO	NRW	STP	CLE	OMC	CPA	WPP	SZE	STR	LOC	PCT
Working Ratio	r	1	0.217	0.111	0.024	0.440	−0.129	0.015	−0.291	−0.154	0.293	−0.317
	Sig.		0.009	0.187	0.778	0.000	0.125	0.855	0.000	0.067	0.000	0.000
	N	143	143	143	143	143	143	143	143	143	143	143
Non Revenue Water	r	0.217	1	0.380	−0.152	0.025	−0.085	0.433	0.278	−0.291	0.127	−0.242
	Sig.	0.009		0.000	0.069	0.768	0.312	0.000	0.001	0.000	0.131	0.004
	N	143	143	143	143	143	143	143	143	143	143	143
Staff Productivity	r	0.111	0.380	1	−0.027	0.073	0.127	0.387	0.236	−0.079	−0.104	−0.008
	Sig.	0.187	0.000		0.749	0.384	0.132	0.000	0.005	0.347	0.218	0.922
	N	143	143	143	143	143	143	143	143	143	143	143
Collection Efficiency	r	0.024	−0.152	−0.027	1	0.320	−0.047	−0.146	−0.017	0.159	−0.460	0.325
	Sig.	0.778	0.069	0.749		0.000	0.581	0.083	0.842	0.058	0.000	0.000
	N	143	143	143	143	143	143	143	143	143	143	143

Continued

Operating & Maintenance Cost	r	0.440	0.025	0.073	0.320	1	−0.494	−0.564	−0.005	0.222	−0.111	0.548
	Sig.	0.000	0.768	0.384	0.000		0.000	0.000	0.956	0.008	0.185	0.000
	N	143	143	143	143	143	143	143	143	143	143	143
Per Capita	r	−0.129	−0.085	0.127	−0.047	−0.494	1	0.394	0.015	−0.337	−0.083	−0.413
	Sig.	0.125	0.312	0.132	0.581	0.000		0.000	0.861	0.000	0.323	0.000
	N	143	143	143	143	143	143	143	143	143	143	143
Production Percent	r	0.015	0.433	0.387	−0.146	−0.564	0.394	1	0.148	−0.292	0.044	−0.543
	Sig.	0.855	0.000	0.000	0.083	0.000	0.000		0.078	0.000	0.605	0.000
	N	143	143	143	143	143	143	143	143	143	143	143
Provider Size	r	−0.291	0.278	0.236	−0.017	−0.005	0.015	0.148	1	0.069	−0.116	0.147
	Sig.	0.000	0.001	0.005	0.842	0.956	0.861	0.078		0.413	0.168	0.079
	N	143	143	143	143	143	143	143	143	143	143	143
Provider Structure	r	−0.154	−0.291	−0.079	0.159	0.222	−0.337	−0.292	0.069	1	−0.253	0.454
	Sig.	0.067	0.000	0.347	0.058	0.008	0.000	0.000	0.413		0.002	0.000
	N	143	143	143	143	143	143	143	143	143	143	143
Provider Location	r	0.293	0.127	−0.104	−0.460	−0.111	−0.083	0.044	−0.116	−0.253	1	−0.442
	Sig.	0.000	0.131	0.218	0.000	0.185	0.323	0.605	0.168	0.002		0.000
	N	143	143	143	143	143	143	143	143	143	143	143
Price Category	r	−0.317	−0.242	−0.008	0.325	0.548	−0.413	−0.543	0.147	0.454	−0.442	1
	Sig.	0.000	0.004	0.922	0.000	0.000	0.000	0.000	0.079	0.000	0.000	
	N	143	143	143	143	143	143	143	143	143	143	143

6

Evaluating and Improving the Effectiveness of Vehicle Inspection and Maintenance Programs: A Cost-Benefit Analysis Framework

Ying Li

Department of Environmental Health, College of Public Health, East Tennessee State University, Johnson City, USA
Email: liy005@etsu.edu

Abstract

Motor vehicle inspection and maintenance (I/M) programs are designed to identify high-emitting vehicles and mitigate their impacts on air quality and climate. I/M programs have been traditionally ranked superior among various vehicle emission control measures by the results of cost-benefit analysis, based on the assumption that these programs will achieve the targeted emission reduction outcomes. However, the actual effects of I/M programs may be greatly uncertain and when this uncertainty is taken into account, these programs may become suboptimal. This study develops a new a cost-benefit analysis framework that links various program design consideration, such as program participation rate, identification rate and effective repair rate, to the public health benefits as well as costs of the programs. This framework helps decision makers to investigate minimum implementation requirements that at least ensure the benefits are greater than the costs of implementing the programs in order to improve the overall effectiveness of the I/M programs. To illustrate the applications of the framework, it was applied to a particulate matter oriented I/M program targeting all diesel-fueled vehicles in the city of Bangkok, Thailand, a large metropolitan area that has been suffering from severe ambient PM pollution mainly attributable to its wide use of diesel-fueled vehicles and motorcycles. It was found that the health benefits achieved from the program are sensitive to several key program design elements, including participation rate and problem vehicle identification rate, fraction of effective repairs and illegal operation rate. Other variables, such as the testing cut-points and vehicle population growth rate, only have modest effects on the overall emission reduction and consequent health benefits. Overall, the performance of multiple variables associated with I/M program design needs to be improved simultaneous in order to achieve the targeted benefits of the program.

Keywords

Inspection and Maintenance Programs, Cost-Benefit Analysis, Effectiveness, Program Design

1. Introduction

Motor vehicle inspection and maintenance (I/M) programs are designed to identify high-emitting vehicles and mitigate their impacts on air quality and climate [1]. In western developed countries such as the United States (US), these programs have been considered to be cost-effective and are required by the Clean Air Act Amendments of 1990 in regions with the most challenging air pollution problems [2] [3]. The main purpose of I/M programs is to encourage better maintenance for in-use vehicles and to assure the vehicle emission control systems are functioning properly through periodic inspections. The rationale for an I/M program is that the emission distribution among a vehicle population is highly skewed: A small portion of vehicles (estimated at 5% - 10%), sometimes called gross polluter vehicles, is responsible for a substantial fraction (variously estimated at 50% to 80%) of total vehicle emissions [4] [5]. Moreover, not only old vehicles can be gross polluters, but also vehicles of all model years may include some proportion of gross polluters [5], due to the factor that vehicle emission levels are heavily dependent on maintenance. This problem can be even more pronounced in developing countries, where vehicles have a long lifetime and are often poorly maintained [6]. In this case, upgrading maintenance practices and replacing the worst engines should be considered first before moving on to better technology [7]. In addition, despite technological and regulatory advances, new vehicle standards are not sufficient to achieve pollution abatement goals if vehicles deteriorate rapidly, resulting in increasing emission rates [6]. Therefore, to control rapidly growing vehicle emissions, governments must not only affect the behavior of vehicle manufacturers and fuel suppliers, but also the actions of drivers in terms of how well they maintain their vehicles regardless of their vehicle ages [8].

However, although simple in concept, the detailed design and implementation of I/M programs is challenging. For example, when emission control equipment malfunctions, vehicle performance may be unaffected, hence the driver has no private incentive to seek repairs, and demanding private expenditures of money and time by vehicle owners will create the usual tensions that lead many actors to try to evade the regulation in numerous ways [8]. The practices of I/M programs in the US have shown various barriers that may cause failure of these programs to generate the emission reduction originally anticipated by policy makers. For instance, motorists may have many opportunities to evade required repairs, such as testing vehicles numerous times until they happen to pass. By and large, launching an effective I/M program requires massive behavior change among the drivers of a region [8].

Due to these challenges, the actual cost-effectiveness of I/M programs often remain uncertain. In theory, vehicular emission reductions available from I/M programs are mainly determined by the failure thresholds, or cut points, used to identify problem vehicles, but are also quite sensitive to a variety of factors, such as the actual percentage of problem vehicles identified, the percentage of problem vehicles waived from repairs or operating illegally, the emission reduction achieved by repairs, the durability of repairs, and so on [2]. In practice, these factors are often overlooked in designing and evaluating a regional I/M program.

This paper aims at developing a cost-benefit analysis framework for evaluating the effectiveness of vehicle inspection and maintenance programs, on the basis of the emission reduction assessment tool called "I/M Design" developed by [2] (referred to as EISINGER2005 hereafter), and the health benefit analysis tool developed in our previous study [9]. This combined analysis framework incorporates various factors that affect the level of emissions achieved by an I/M program, and links emission reduction with health benefits (avoided mortality and illnesses). The new framework is then applied to a hypothetical particulate matter (PM) oriented I/M program targeting all diesel-fueled vehicles in the city of Bangkok, Thailand, a megacity that has been suffering from severe adverse health effects attributable to PM for a few decades, to illustrate how the framework may help to evaluate the cost-effectiveness of I/M programs.

2. Methods and Data

2.1. Theoretical Framework

I/M programs are one of the essential policy tools to control emissions from in-service vehicles in severely polluted urban areas [2]. An I/M program has the potential to reduce emissions in a number of ways, such as better maintenance of vehicles by motorists as a result of the program, repairs made in anticipation of an I/M inspection (referred to as pre-test repairs) or as a result of failing the test, and early scrapping of vehicles that are not worth repairing [3]. **Figure 1** illustrates the sources of emissions reductions resulting from an I/M program.

Based on this conceptual framework, EISINGER2005 developed a spreadsheet tool to evaluate the effectiveness of I/M programs in terms of levels of emission reductions achieved by those programs. The theoretical basis of this tool is that the amount of vehicle emission reductions resulting from an I/M program is a function of the following variables [2]:

- Pre-I/M test repair work: potential vehicle repairs motivated by instituting I/M for the vehicles that would otherwise be identified as problems.
- Post-I/M test repair work: emission reductions from repair work for the vehicles identified as problems by inspections. This variable itself is a function of the number (or the percentage) of the problem vehicles identified by I/M and the number (or the percentage) of identified problem vehicles being repaired effectively.

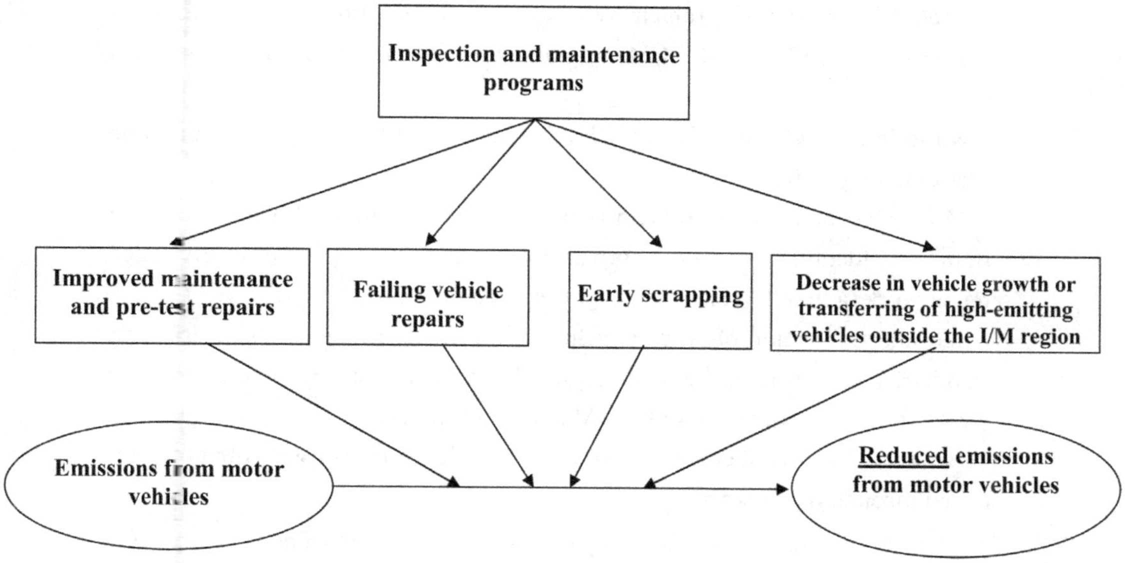

Figure 1. Sources of emission reductions from I/M programs (Partial information source: [3]).

- Vehicles scrapped: emission reductions will be generated from scrapping gross polluting vehicles that would otherwise continue to be used, and replacing them by low emission vehicles. This variable also includes high-polluting vehicles that are transferred outside the I/M region due to the implementation of the programs.

Based on this theoretical basis, this study modified the equations developed by EISINGER2005 and included six equations in the Spreadsheet used to estimate the benefits and costs of an I/M program as follows (all six equations Equations (1)-(6) were originally developed by EISINGER2005, and adopted in this study with minor modifications):

Equation (1) describes the percentage of all problem vehicles that are identified by an I/M program. This variable is a function of the program participation rate and problem vehicle identification rate:

$$\text{ProbVeh} = \text{PartiRate} \times \text{IndenRate} \tag{1}$$

where:

ProbVeh: Percent of all problem vehicles that are identified by an I/M program.

PartiRate: Percent of total required vehicles that participate in an I/M program.

IndenRate: Percent of problem vehicles inspected that fail the test.

Equation (2) describes the percentage of problem vehicles that are both identified by I/M and subsequently undergo repair work:

$$\text{PercentRep} = \text{ProbVeh} \times \left[(1-\text{ScrapFrac}) \times (1-\text{VehWaive}) \times (1-\text{IllegalVeh})\right] \tag{2}$$

where:

PercentRep: Percent of all problem vehicles that are failed (identified) by I/M and subsequently repaired.

ProbVeh: Fraction of problem vehicles (vehicles emitting above certification standards) identified by the I/M program test (this is a function of false pass rates).

ScrapFrac: Percent of failed vehicles that are retired from the fleet within one year of failing their I/M test.

VehWaive: Fraction of problem vehicles (vehicles emitting above certification standards) identified by the I/M program test but allowed a waiver from needed repair work (a function of money spent on repairs).

IllegalVeh: Fraction of problem vehicles (vehicles emitting above certification standards) operating without obtaining the requisite repairs or certifications needed to pass or waive out of the I/M inspection process.

Equation (3) describes the percent of a vehicle's total emissions that are reduced through repair work:

$$\text{PercentRed} = \text{GoodRep} \times \text{ExEm} \times \text{DurRep} \times \text{EmisFrac} \tag{3}$$

where:

PercentRed: Percent of total vehicle emission reductions achieved by repairs, for the vehicles failing I/M and getting repaired (does not include vehicles that fail I/M and are scrapped, waived, or illegally operating).

GoodRep: Fraction of repairs that are "good" (effective), as measured by percent of repaired vehicles that immediately pass a retest.

ExEm: Fraction of excess emissions (where "excess" means emissions above allowable levels, usually referred to as the "cutpoint") reduced from identified problem vehicles that receive good repairs (I/M does not address all excess emissions, for example cold start emission problems).

DurRep: Durability of good repairs, as measured by percent of vehicles with good repairs that pass retests at 12 or 24 months.

EmisFrac: Fraction of total vehicle emissions represented by pre-repair excess emissions (this is a function of the "cutpoint" used to define the point at which a vehicle is allowed to pass I/M; emissions above the passing cutpoint are considered excess). In other words, emissions below I/M cutpoints are essentially acceptable, emissions above cutpoints are excess; this variable represents the percent of total vehicle emissions considered excess.

Equation (4) describes the benefits of repair work after I/M test:

$$\text{BenefitsRep} = \text{PercentRep} \times \text{PercentRed} \tag{4}$$

where:

BenefitsRep: The percentage of vehicles repaired (PercentRep), multiplied by the percentage reduction achieved per repair (PercentRed); units are in percent of total vehicle emissions reduced.

Equation (5) describes emission reduction benefits from vehicle retirements due to I/M test failures:

$$\text{BenefitsScrap} = \text{ProbVeh} \times \text{ScrapFrac} \times \text{ScrapEmis} \tag{5}$$

where:

BenefitsScrap: Percent reductions from all problem vehicles, due to vehicles that are scrapped (considering emissions from the replacement vehicles).

ScrapEmis: Percent of total vehicle emissions reduced, for each vehicle retired from the fleet, after accounting for replacement vehicle emissions.

Equation (6) describes total program benefits, in terms of the percent emission reduction in total vehicle emissions from the I/M program:

$$\text{BenefitsTotal} = \text{BenefitsGrow} + \text{BenefitsRep} + \text{BenefitsScrap} \tag{6}$$

where:

BenefitsGrow: Percent emission reductions achieved due to the changes in vehicle population growth as a result of I/M enforcement. It is calculated by using the projected total number of vehicles under the baseline and the I/M scenarios, as well as the fleet-average emission rates. This variable is not in the original "I/M Design" spreadsheet, but is developed by this study to reflect future I/M effectiveness as vehicle fleets change over time.

The unit for BenefitsTotal is percent of total vehicle emissions reduced, for the entire problem vehicle fleet.

This spreadsheet tool, developed by modifying the tool given in EISINGER2005, allows users to adjust the values of parameters in the model and obtain the resulting percentage of emission reductions in total vehicle emissions. Although the tool is developed in the US for the most common gasoline vehicle I/M programs, the fundamental ideas of vehicle I/M programs are universal and thus the theoretical modeling framework applies to I/M programs targeting other pollutants in other regions. However, the values of the variables must reflect the specific contexts and issues of concern with respect to the interested area and programs[1].

The present study used this spreadsheet tool as a cost-benefits framework that links I/M design considerations with health benefits associated with the programs in order to understand the impacts of some key issues regarding I/M design, such as compliance rates, testing cut-points and effectiveness of repairs, on the potential health benefits of the programs.

Our previous study estimated the potential health benefits associated with the proposed PM-oriented I/M programs targeting all diesel-fueled vehicles and motorcycles in the city of Bangkok, Thailand, a megacity in Asia that has been suffering from severe adverse health effects attributable to ambient PM for decades [9]. In that study, the health benefits as a function of different levels of PM_{10} (all particulates with an aerodynamic diameter of less than or equal to 10 μm) emission reductions were analyzed and compared with the social costs of the I/M programs (these different levels of reductions were considered due to the significant uncertainty involved in the actual emission reduction benefits of I/M programs). It was found that a minimum of about 4% reduction of the total PM_{10} emissions from motor vehicles is required in order for the total benefits to

[1]Based on a conversation with the author of the paper, Dr. Douglas S. Eisinger.

be greater than the total costs of implementing the programs. On the basis of that study, the main purpose of the current study is to examine how key variables affect PM_{10} emission reductions available from the same I/M programs, and the desirable performance of these variables in order to achieve the 4% minimum emission reduction objective. The best available information about I/M experience in Thailand and elsewhere were used as the inputs to the models in spreadsheets.

2.2. A Framework to Estimate the Effectiveness of I/M Programs

Empirical evidence on the performance of I/M programs and on the important elements affecting I/M emission reductions is fairly limited. Evidence on the performance of PM-oriented I/M is even less available given that these programs are still relatively new. In Thailand, the data collected by a World Bank study [10] on their pilot motorcycle inspection and upgrade project in Bangkok are the most comprehensive dataset on I/M programs in the Bangkok Metropolitan Region (BMR). Very little information is available on the performance of I/M targeting diesel-fuel vehicles including buses and trucks. Given the limitation of data, in running the spreadsheet developed by EISINGER2005, the values of most variables in the tool are derived based on the best available information in the U.S. and some extrapolation is performed to the BMR. Definitions of the variables in "I/M Design" and their input values used in this study are presented in **Table 1**.

In addition to the input variables listed in **Table 1**, the spreadsheet also needs the following inputs related to the characteristics of the vehicle population studied: 1) Problem vehicles as percent of total vehicles: The values used in the spreadsheet were consistent with the assumptions made in our previous study—10% of buses and heavy trucks, 17.5% of light trucks and 25% of motorcycles in the BMR are problem vehicles [9]. In the uncertainty analysis, the upper and lower bounds of this parameter were assumed to be 1.5 times and half of the mean estimate, respectively. In lack of empirical evidence to support the form of PDF of this parameter, the uniform distribution was selected based on the authors own judgment; 2) Problem vehicles as percent of total PM emissions: Studies usually suggest that the gross-polluting vehicle pool is responsible for a substantial fraction—ariously estimated at 50% to 80%—of total vehicle emissions [4] [5]. Based on this, this study assumes that 50% and 80% are the lower and upper limits, respectively, and the mean value of them, 65%, is the best estimate of total vehicular emissions are generated by problem vehicles. Also, the uniform distribution was selected in the uncertainty analysis based on the author's own judgment; 3) Number of vehicles (under both the baseline and the I/M scenarios), average annual vehicle kilometers traveled (VKT) per vehicle and baseline fleet-average emission rates (in the unit of g/km-vehicle).

The testing cut-points for each type of vehicle need to be determined and input into the spreadsheet. Emission cut-points are established in I/M programs to

Table 1 Variables governing I/M program emission reductions (All variables in this table were developed and defined in EISINGER2005 unless noted).

Variable name and definition	Variable explanation	Values used in EISINGER2005[1]	Values used in this study (Range, PDF)	Discussion
PartiRate: % of all vehicles required by I/M participate in the programs	Although I/M programs require all vehicles regulated by the programs to take the inspection process, there may be a certain fraction of vehicles operating illegally without participating in I/M. This variable is not in the "I/M Design" spreadsheet[2] PT, but is developed by this study to reflect the levels of participation in the programs in the study area.	100%	90% (80% - 100%, triangular[3])	It has been proposed that the new PM-related I/M programs should be linked to vehicle registrations and managed by a central database in order to significantly improve the levels of participation in the programs. It is expected that, with the government's strong will and efforts to curb severe air pollution in the BMR, the participation rate of the programs can be high. A mean estimate of 90% participation rate is assumed in this study. However, sensitivity analysis will test the role of this variable on the overall emission reductions by I/M. For simplicity, it was assumed that the fraction of problem vehicles is the same in the participation group as in the "non-participation group", although in reality, problem vehicles are more likely to escape from the inspection process.
IndenRate: % of inspected problem vehicles that are identified by I/M	Although I/M programs aim at identifying all the problem vehicles (defined as vehicles whose emission rates exceeds I/M testing cut-points) that are inspected, the inherent limitations of I/M make a 100% identification rate unrealistic. It is accepted that some problem vehicles, e.g. 10% of all problem vehicles, will falsely pass I/M.	Upper: 90% Lower: 71%	50% (0% - 100%, triangular)	This variable reflects the ability of I/M programs to identify problem vehicles. Given that at present the test protocol and program implementation for PM-related I/M programs is not as well developed as that for traditional I/M, and both may be less well developed in developing countries, the identification rates associated with these I/M programs are expected to be lower. Based on this, a 50% identification rate is assumed in this study. In uncertainty analysis, the range of this variable is set to be 0% - 100%, reflecting the worst case that none of the problem vehicles are identified and the ideal case that all problem vehicles are identified.

Continued

ScrapFrac: % of failed vehicles that are scrapped	Early scrappage of problem vehicles results in emissions reduction, if the replacement vehicles generate less emission. For simplification, "I/M Design" does not account for replacement vehicle deterioration, and all replacement vehicles are assumed to pass I/M two years following their purchase.	Upper: 13.3% Lower: 6.7%	5% (0% - 10%) for motorcycles, 2.5% (0% - 5%) for light duty trucks, 0.5% (0% - 1%) for buses and heavy trucks (Triangular distribution)	A World Bank study [10] predicts that 5% of failed motorcycles will be scrapped. This result indicates that the early scrappage rates attributable to I/M programs may be lower in developing countries than in developed countries. 5% is applied to motorcycles in this study, and a 2.5% scrappage rate is assumed for light-duty trucks and 0.5% for buses and heavy trucks, given that these vehicles are generally more expensive and thus less likely to be scrapped.
VehWaive: % of identified problem vehicles waived by an I/M program	I/M design in the U.S. generally allows some fraction of problem vehicles to be waived from the programs, usually because of economic hardship.	Upper: 1% Lower: 4%	1% (0-2%, uniform)	In Bangkok, the government may also consider waivers in the implementation of I/M programs. In particular, for public transit such as buses, high repair costs are likely to result in significant increases in bus fares, which may prevent low income people from using them. However, given the severity of the air pollution problem in the area, a high waiver rate should be restricted. A 1% waiver rate was assumed for all vehicle types.

Continued

IllegalVeh: % of identified problem vehicles operate illegally	There may be some fraction of vehicles operating illegally without undergoing the requisite repairs or certifications needed to pass or be waived from the I/M inspection process.	Upper: 6.6% Lower: 13%	Motorcycles and light trucks: 20% (0% - 40%); buses and heavy trucks: 10% (0% - 20%) (Triangular distribution)	The illegal operating rates may be higher in the BMR since the I/M programs are less mature. It was assumed that for motorcycles and light trucks, the rates are both 20%, and for public transits and heavy trucks, the rates are 10% since it should be easier to identify the violations by these vehicles on road.
GoodRep: % of repair work initially effective	Some fraction of repairs are not effective but falsely pass re-tests. For example, random roadside tests show that a portion of the vehicle fleet fails I/M immediately after being repaired but then pass an official I/M test.	Upper and Lower: 80%	72% (44% - 100%, triangular)	Available information related to this variable is very limited. A study by Land Transport Department of Thailand randomly selected 21 private inspection centers in Bangkok and requested two problem motorcycles to be tested by these inspection stations [7]. The two motorcycles were failed by 12 of the 21 stations whereas passed by the remaining 9 stations [7]. This study indicates that only 58% $(1-9\div21)$ of testing vehicles may properly pass the I/M. It was assumed that the updated I/M in the BMR considered here will improve the performance of this variable and achieve 90% of the US level. Therefore, the value of this variable is: $90\%\times80\%=72\%$.

Continued

ExEm: % of excess emissions (emissions above allowable levels) from identified problem vehicle reduced by good repairs (repairs that properly pass an I/M test immediately)	Effective repairs motivated by I/M do not address all excess emissions. For example, I/M tests do not address emissions from cold starts, since vehicles are tested after the engine and catalyst are warm. US EPA estimated that a model IM240 program identifies 92% of HC, 68% of carbon monoxide (CO), and 83% of nitrogen oxides (NO_x) excess emissions.	Upper: 92% Lower: 81%	81% (62% - 100%, triangular)	This variable is highly uncertain for I/M programs targeting PM without further research. In lack of further information, the rates of the three pollutants HC (92%), CO (68%), and NO_x (83%) are averaged (equal to 81%) and used for PM. The range is 62% - 100%.
DurRep: % of good repairs that remain durable	Some of the good repairs may deteriorate fast and not be durable enough to pass another I/M test after one or two years (depending on the frequency of testing required). Therefore, they will generate excess emissions in between two tests.	Upper: 94% Lower: 79%	86.5% (73% - 100%, triangular)	Diesel vehicles may deteriorate rapidly without proper maintenance. On the contrary, a well-maintained diesel vehicle will generally retain a good emissions performance throughout its operating life [11]. It is expected that the updated I/M programs in the BMR should be able to motivate vehicle owners to better maintain their vehicles in anticipation of the effective inspection process. In lack of more available information, the US values were used in this study.

[1]The input values used in EISINGER2005 were based on light-duty vehicle hydrocarbon (HC) inspection data from an enhanced I/M program in southern California's South Coast Air Basin. [2]The original study examines the emission reduction benefits by an I/M program in a previous year using data on the actual number of vehicles inspected. It assumed that all vehicles subject to inspection participated in the program. [3]Triangular distribution was selected for all the variables in **Table 1** except for VehWaive. There is little empirical evidence to support the PDFs of the variables in **Table 1**. Therefore, the PDFs were selected based on the author's own judgment. When there is more confidence in values near the central value than in values far away on either side, the triangular distribution was selected. In the case of the variable VehWaive, there is no reason to believe that some values between the lower and upper limits are larger than the others, and therefore, the uniform distribution was selected. Furthermore, the range was obtained based on the mean and the known theoretical limit of the variable, namely, 0% for the lower limit or 100% for the upper limit.

identify the worst polluters and minimize false failures [2]. In reality, vehicle emission rates usually span a wide spectrum. Conceptually, if an I/M targets a 25% reduction of the total emissions from motor vehicles, a cut-point equal to 75% of the current fleet-average emission rate will reduce the emission rates of all vehicles to 75% of the current level or lower, and thus ensure that the 25% emission reduction target is reached with confidence. However, given that the emission rates of the large portion of "good" vehicles (e.g. 90% of total vehicles) are usually much lower than the small portion of problems vehicles (e.g. 10% of total vehicles), it is not necessary, or probably not feasible either, to cut the emission rates of all vehicles to 75% of the current average level or lower in order to achieve the 25% reduction goal[2]. More stringent cut-points may be able to fail more vehicles, in particular those with emission rates close to the failure cut-points. However, more stringent cut-points are also likely to increase the social costs of I/M programs, and to suffer from problems such as technological infeasibility and motorist acceptance of the programs. While how to select and modify testing cut-points in I/M design to optimize the program effectiveness is beyond the scope of this study, this study uses the "ideal" cut-points discussed above, *i.e.* cut-points equal to 75% of the baseline fleet-average emission rates for each vehicle type in the BMR for the "best estimate" case, followed by an examination of the impacts of alternative cut-points on overall emission reduction levels. Just for comparison, the enhanced I/M in southern California's South Coast Air Basin studied in EISINGER2005 used a rate of 86% of the baseline the fleet-average emission rate as the failure cut-point (the baseline rate was 1.25 g/mi and the cut-point was 1.08 g/mi). **Table 2** summaries the fleet-average

Table 2. Fleet-average PM_{10} emission rate in the BMR under the baseline and I/M Scenario (Year: 2008).

Vehicle Type	Fleet-average PM_{10} emission rate (g/km)	
	Baseline[1]	I/M[2]
City Bus	1.231	0.923
City Truck	1.231	0.923
Long Haul Truck/Bus	1.231	0.923
Light Duty Truck	0.264	0.198
Passenger Car	0.003	0.002
Motorcycle	0.100	0.075

Notes: [1]The baseline emission rates were derived from data published in [9] using PM_{10} emission rates in the year 2000 and the assumption that a 5% annual PM emission factor decrease rate for all types of vehicles in the BMR. [2]The emission rates under the I/M scenario were calculated as baseline rate × 75%, assuming cut-points equal to 75% of the baseline fleet-average emission rates for each vehicle type.

[2]A hypothetical example is provided here: Assuming that a vehicle fleet has an average emission rate of 1.0 g/km. 10% of all vehicles are gross polluting and they are responsible for 50% of the total emissions. Based on this information, it can be derived that the average emission rates for good and problem vehicles are 0.56 g/km and 5.0 g/km, respectively. As long as the average emission rate for problem vehicles goes down to 2.46 g/km, the fleet-average rate will decrease to 0.75 g/km. Therefore, if all problem vehicles can be properly identified and fixed, a cut-point of 2.46 g/km will ensure the 25% emission reduction goal accomplished.

PM_{10} emission rates for various vehicle type under the baseline scenario and the I/M scenario.

EISINGER2005 considered that a small fraction (in the range of 0% - 7.5%) of the initial problem vehicles seek repairs in anticipation of I/M tests, and they were assumed to be "good" vehicles in inspection and pass the I/M test. Although the emission reductions resulting from this kind of "pre-test" repairs were taken into account in EISINGER2005, these reductions are only responsible for a small fraction of total emission reduction benefits achieved by I/M programs, approximately ranging from 0% - 2%. It is expected that the fraction of problem vehicles seeking emission repairs before I/M will be even smaller in a developing country than that in the US, given that people are generally less wealthy and less able to afford the costs of maintenance and repairs. For simplicity, this study did not consider the emission reductions resulting from pre-test repairs.

3. Results and Discussion

3.1. Estimating Emission Reduction Effectiveness of the PM-Related I/M Programs in the BMR

Using the "best estimate" values of the variables listed in **Table 1** and the estimated 2008 vehicle population, emission rates and VKT [9], the "I/M Design" spreadsheet was run for the year 2008. The results show that the PM-oriented I/M programs are expected to reduce total PM_{10} emissions from motor vehicles in the BMR by 10.6%. **Table 3** summarizes the findings.

Therefore, in the "best estimate" case, the proposed PM-oriented I/M programs in the BMR are expected to yield health benefits that exceed the social costs of the programs (the "threshold" for achieving this goal is a 4% overall PM emission reduction achieved by the programs, as found in [9].

3.2. Examining the Roles of Key Design Elements on the Emission Reduction Benefits of I/M Programs

A series of sensitivity analyses were conducted to examine the sensitivity of the effectiveness of the I/M programs in terms of the percent of overall PM_{10} emission reduction to the key design elements.

3.2.1. The Effects of Testing Cut-Points on Overall Emissions Benefits

The "best-estimate" in Section 3.1 is based on the assumption that failure cut-points are 75% of the baseline emission rates for each vehicle type. Since the cut-points determine the size of the initial problem vehicle pool (a more stringent testing cut-point is likely to result in more vehicles with "excess" emissions and, hence, subject to repair or replacement), changes to the cut-points will result in changes in the other two inputs: problem vehicles as percent of total vehicles and problem vehicles as percent of total emissions. It is difficult to estimate the magnitude of changes in these two variables as a result of the changes in failure cut-points without knowing the distribution of emission rates. Here it

Table 3. Estimated PM_{10} emission reduction benefits of PM-related I/M programs in the bangkok metropolitan region.

Vehicle Type / Variables	City Bus	City Truck	Long Haul Truck/Bus	Light Duty Truck	Motorcycle	**Total**
Percent of all problem vehicles indentified by I/M (ProbVeh)	45%	45%	45%	45%	45%	
Percent of problem vehicles failed and repaired (PercentRep)	40%	40%	40%	35%	34%	
Percent reductions from all problem vehicles, achieved by post-test repairs (PercentRed)	45%	45%	45%	40%	36%	
Percent reductions due to the decrease in vehicle growth (BenefitsGrow)	0.1%	0.3%	0.3%	1.4%	0.6%	
Benefits of post-test repair wok (BenefitsRep)	17.8%	17.8%	17.8%	13.9%	12.1%	
Percent reductions from all problem vehicles due to scrap (BenefitsScrap)	0.2%	0.2%	0.2%	0.9%	1.6%	
Total reductions, as percent of total emissions from all problem vehicles (BenefitsTotal)	18.1%	18.3%	18.3%	16.2%	14.3%	
Total reductions achieved by each type of vehicles (in tonnes per year)	385	210	75	1253	258	**2180**
Percent reductions achieved within each type of vehicles	11.8%	11.9%	11.9%	10.4%	9.2%	**10.6%**

was assumed that slight changes in cut-points do not change the values of the two variables (this may be true in the case that the majority of good vehicles have emission rates much lower than the cut-points, and the majority of problem vehicles have emission rates much higher than the cut-points), so in this case failure cut-points only affect the new emission rates of problem vehicles after repairs and retests. Based on this assumption, a cut-points sensitivity analysis was conducted. **Figure 2** shows the results.

Figure 2 was generated based on the assumption that all the other input variables are independent of the failure cut-points, *i.e.* changing the cut-points while holding all other variables constant to examine the sensitivity of overall PM emission reductions to failure cut-points. **Figure 2** indicates that the when cut-points decrease from 100% to 60% of the baseline emission rates, the percent of overall emission reduction increases from 9.8% to 11.1%. Therefore, cut-points modifications within a certain range (e.g. from 60% - 100% of the

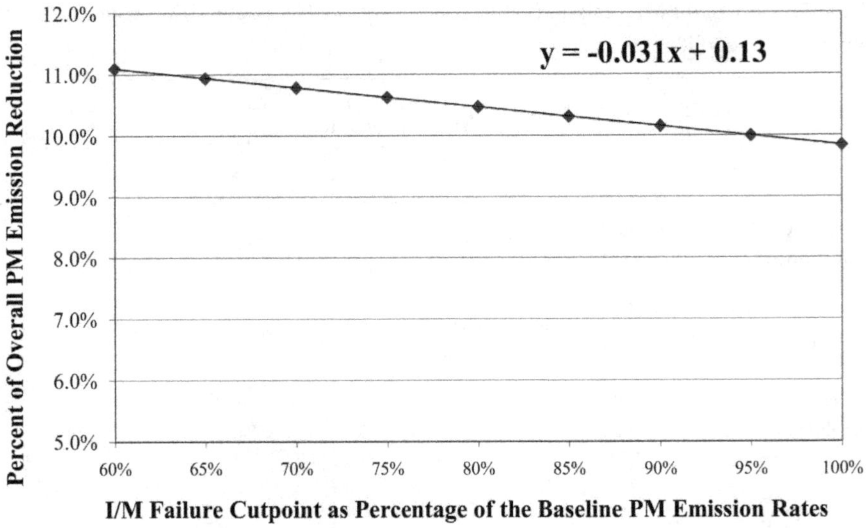

Figure 2. Effects of testing cut-points on the percentage of overall emission reduction by I/M programs.

baseline emission rates in this case) only have modest effects on the overall emission reduction benefits (the percent of overall PM_{10} emission reductions changes from approximately 10% to 11% as the result of changing the failure cut-points from 60% - 100% of the baseline emission rates), because most vehicles are considered to emit at levels well outside the range (either higher or lower the cut-points). This study considers that other I/M design elements may have more significant effects on the emission reduction benefits achieved by I/M programs. In general, important variables in I/M design that policy makers need to address include program participation rate, problem vehicle identification rate, effective emission repair rate and problem vehicle illegal operation rate. In the following section, the impacts of these variables on emission reduction were analyzed.

3.2.2. Key Variables Affecting I/M Effectiveness

1) Participation rate and problem vehicle identification rate associated with I/M programs. Participation rate (PartiRate) and problem vehicle identification rate (IdenRate) are two key elements to address in designing I/M programs. Participation rate represents the levels of program enforcement. A successful I/M program minimizes vehicle violations (vehicles required by an I/M program do not participate in the program). The "best estimate" case in Section 3.1 assumes the majority (90%) of vehicles in the BMR required by I/M will participate in the programs, *i.e.* they will undertake appropriate emission inspection (reasons discussed in **Table 1**). However, if a large fraction of vehicles subject to I/M tests escape from the inspection process, the emission reduction benefits of I/M are expected to decrease considerably. Problem vehicle identification rate represents the ability of I/M programs to identify vehicles that exceed the emission standards and thus need emission repairs.

Sensitivity tests of each of the two variables were conducted by changing the input value of one variable while holding all other inputs constant (presuming that all input variables are independent of each other). The goal is to examine the sensitivity of the percent of overall emission reduction to program participation rate (PartiRate) or problem vehicle identification rate (IdenRate). **Figure 3** shows the results.

The square-marked and the triangle-marked lines represent the percent of overall PM emission reductions achieved by I/M programs as a function of program participation rate (PartiRate) and problem vehicle identification rate (IdenRate), respectively (as noted in the figure). Each of the two lines was generated by incrementing the value of an individual variable (PartiRate or IdenRate) by 10% at a time (starting from 0% and ending at 100%), while setting all other inputs to their best estimates. And the red solid line represents the minimum percentage of emission reduction required in order for the benefits of the programs to outweigh the costs (the value was 4% as found in [9]).

Figure 3 indicates that both participation rate and problem vehicle identification rate are important determinants of overall PM emission reduction benefits achieved by I/M programs. For participation rate, when the value of this variable increases from 0% (lower bound) to 100% (upper bound), the percent of overall emission reductions from vehicles increases from 0.6% to 11.7%; for problem vehicle identification rate, the percent of overall emission reductions from vehicles increases from 0.6% to 20.6% when the variable's value changes from the lowest to the highest. Comparing the effects of the two variables in **Figure 3** shows that problem vehicle identification rate has a greater impacts on the overall emission reduction benefits than program participation rate, since for the same increment (e.g. 10%) in the two variables, the incremental emission reduction benefits resulting from the change in the problem vehicle identification rate

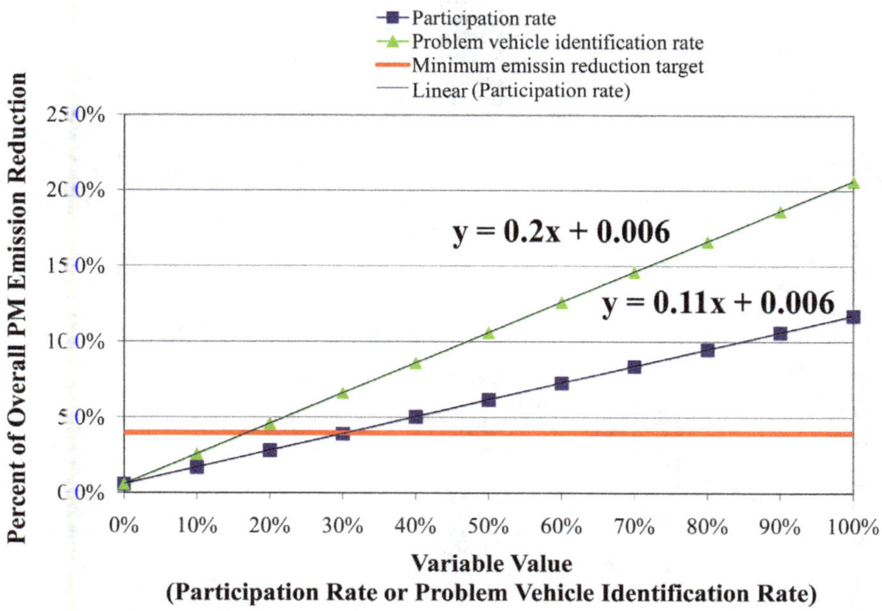

Figure 3. Effects of participation rate and problem vehicle identification rate on the percent of overall PM emission reduction by I/M programs.

are greater.

In order to achieve the goal of 4% PM emission reduction from motor vehicles, the participation rate is required to be greater than 30%, if all other inputs remain the same values as in the "best estimate" case. And the requirement for problem vehicle identification rate is 17% when setting the other variables in the spreadsheet to their "best estimate" values.

2) The impacts of the effectiveness of problem vehicle repairs. Repairing problem vehicles to meet emission standards is the major source of emission reduction available from I/M programs. Three variables in the I/M Design spreadsheet are related to the effectiveness of repairs: GoodRep—Percent of repair work initially effective; ExEm—Percent of excess emissions (emissions above allowable levels) from identified problem vehicles reduced by repairs that properly pass an I/M test immediately; and DurRep—Percent of good repairs that remain durable until the next I/M inspection. **Figure 4** shows the impact of each individual variable on the levels of emission reduction achieved by I/M programs.

The three marked lines (named as GoodRep, ExEm and DurRep) were generated using the same approach as used to generate **Figure 3**: Each line is generated by incrementing the value of the individual variable it represents (GoodRep, ExEm or DurRep) by 10% at a time (starting from 0% and ending at 100%), while setting the other inputs to their "best estimate" values. The red solid line also represents the minimum PM emission reduction target of 4%.

Figure 4 illustrates that the increase in the values of any of the three variables related to problem vehicle repairs results in considerable improvement in emission reduction performance by the I/M programs, as these variables are key

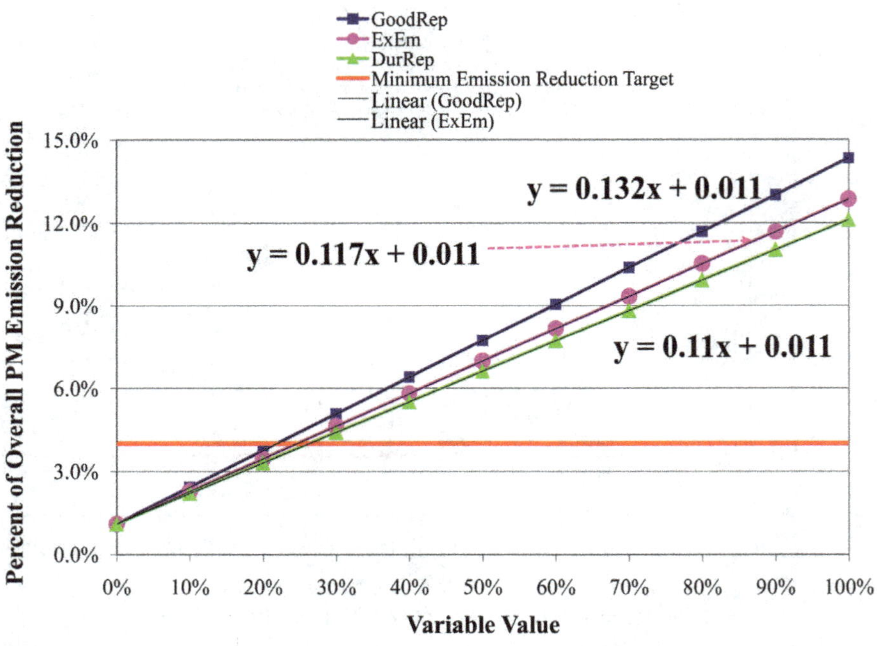

Figure 4. Effects of problem vehicle repairs on the percent of overall PM emission reduction by I/M programs.

determinants of the effectiveness of emission repairs. Comparing the three marked lines in **Figure 4** shows that while the same increment in any one of the three variables results in approximately similar incremental emission reduction benefits, the impact of initially effective repair rate (GoodRep) is slightly greater than the impacts of the other two variables. The 4% emission reduction target requires a minimum of 22% of repair work initially effective (GoodRep), or 25% of excess emissions from identified problem vehicles reduced by repairs (ExEm), or 26% of repairs that properly pass an I/M remain durable until the next I/M inspection (DurRep).

3) The impacts of illegal operation by problem vehicles. Illegal operation here refers specifically to failed vehicles that continue to run on roads without appropriate repairs or certificates of waiver (the variable IllegalVeh in **Table 1**). There are other types of illegal operation in I/M program implementation. For example, vehicles may run on roads without taking the inspection required by the programs. This latter type of illegal operation is considered in the program participation rate variable, so it is not taken into account here. Illegal operation by failed vehicles may considerably damage the performance of I/M programs, since these vehicles are identified as gross emitters. Using the same sensitivity test approach as in **Figures 3-5** was generated, which shows the effects of failed vehicle illegal operation rate on the levels of overall emission reduction achieved by I/M programs.

Therefore, the increases in illegal operation rate by failed problem vehicles can substantially reduce the emission reduction benefits achieved by I/M programs. In order to achieve the goal of 4% PM_{10} emission reduction from motor vehicles, the rate of failed problem vehicle illegal operation should not go over 75%, presuming that the performance of the other variables is at the level of the "best

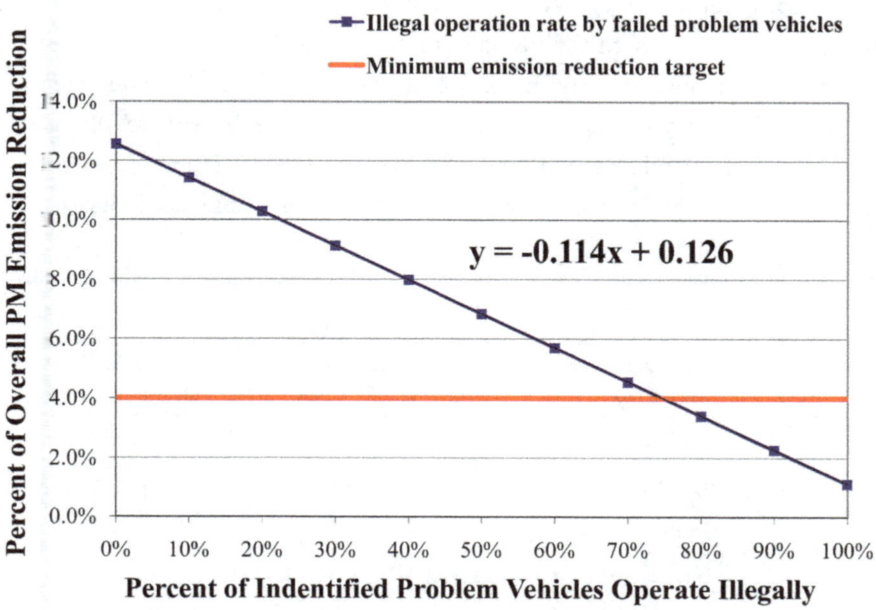

Figure 5. Effects of problem vehicle illegal operation on overall emission reduction benefits.

estimate". In reality, it may be unrealistic to achieve 0% illegal operation by failed problem vehicles. However, minimizing the problem vehicle illegal operation rate is an essential I/M design element to improve the effectiveness of I/M programs.

3.3. Improving the Emission Reduction Effectiveness of I/M Programs

For the PM-oriented I/M programs in this study, a 25% PM_{10} emission reduction in the BMR was originally proposed as an upper bound target of the I/M programs based on the past experience of similar programs in the US [9]. Also as discussed earlier, the levels of PM_{10} emission reductions actually achieved by the programs are significantly uncertain. The results in **Figures 3-5** indicate that based on the assumptions made in **Table 1**, the improvement in the performance of any individual element is not sufficient to achieve the upper bound target of 25% overall emission reduction initially expected in proposing the programs to be adopted in the BMR.

For example, when one of the key variables discussed above reaches the upper bound, *i.e.* 100% (for IllegalVeh, the upper bound is 0%), while holding the other variables the same as in the "best estimate" case, the percent of overall PM_{10} emission reduction is summarized in **Table 3**.

The results in **Table 4** show in the case that only one variable in the spreadsheet increases while the values of the others remain the same as assumed in **Table 1**, even if the variables achieve complete success, the maximum level of PM emission reduction benefits is 20.6% (when IndenRate reaches 100%). Therefore, the performance of more variables needs to be improved simultaneously.

As two illustrations, when the values of PartiRate, IndenRate, IllegalVeh, GoodRep, ExEm and IllegalVeh were replaced by the lower levels found in the I/M program in southern California's South Coast Air Basin [2], the percent of PM emission reduction from motor vehicles increased to 19.1%; and when the values of all the key variables discussed above (PartiRate, IndenRate, IllegalVeh, GoodRep, ExEm, and DurRep) were set to the upper values found in California's I/M, the percent of PM reduction increased to 30.9%. **Table 5** summarizes the

Table 4. Emission reduction benefits in the case that one key variable reaches the upper bound.

Variable reaching the upper bound (100%)	Percent of overall PM_{10} emission reduction from Vehicles
PartiRate	11.7%
IndenRate	20.6%
GoodRep	14.3%
ExEm	12.9%
DurRep	12.1%
IllegalVeh	11.5%

Table 5. Improving the effectiveness of the I/M Programs in the BMR by increasing the values associated with key design elements.

Variable	Best estimates in this study	I/M in Southern California's South Coast Air Basin	
		Lower value	Upper value
PartiRate	90%	100%	100%
IndenRate	50%	71%	90%
GoodRep	72%	80%	80%
ExEm	81%	81%	92%
DurRep	86.5%	(79%)*	94%
IllegalVeh	10% for buses and heavy trucks, 20% for light trucks and motorcycles	13%	6.6%
Percent of overall PM_{10} emission reduction by I/M	**10.6%**	**19.1%**	**30.9%**

**Note*: Data in parentheses were not used in the calculation since they are smaller than the "best estimate" in this study.

replaced variable values used and the new emission reduction estimates.

3.4. Emission Reduction Benefits Due to the Change in Vehicle Population Growth

The "best estimate" in **Table 3** was based on the assumption that the implementation of the new PM-related I/M programs causes 10% decrease in average annual vehicle growth rate in the BMR [9]. This assumption is associated with the variable BenefitsGrow (the percent of emission reductions achieved due to the changes in vehicle population growth as a result of I/M enforcement). Sensitivity test was conducted by changing the percentage decrease in annual vehicle growth rate from 10% to 0% (no change in the annual growth rate), 20% or 30%, while holding all other input unchanged. **Figure 6** summarizes the results.

Past experience in rapidly developing metropolitan areas in Asia shows that the introduction of vehicle I/M programs may slightly slow down the fast growth of motor vehicles in these areas and it is expected that the percent decrease in average annual vehicle growth rate falls into the range of 0% - 30%. **Figure 6** indicates that the change in the assumption about the percent decrease in annual vehicle growth rate has modest impact on the overall emission reduction benefits, when the change falls into the range of 0% - 30%.

3.5. Uncertainty Analysis Results

Contribution to variance is a measure of the fraction of the total uncertainty (variance) in the risk estimate that comes from the uncertainty in a particular parameter, when all parameters are allowed to vary simultaneously [12]. Research to reduce uncertainty then should focus limited resources on narrowing the uncertainty in the premise showing the greatest contribution to variance [12]. **Ta-**

ble 6 summarizes the top 10 variables that contribute the most significantly to the uncertainty in the percent of overall PM_{10} emission reductions. The analysis was conducted using Monte Carlo simulation based on the variable PDFs listed in **Table 1**, and were performed in the Oracle Crystal Ball software. The sample size was set as 5000.

Table 6 indicates that the problem vehicle identification rate of light-duty trucks showed the greatest contribution to variance. The contribution to variance of this premise is large probably both because the overall PM_{10} emission reductions are sensitive to this variable (as found in **Figure 3**), and because the

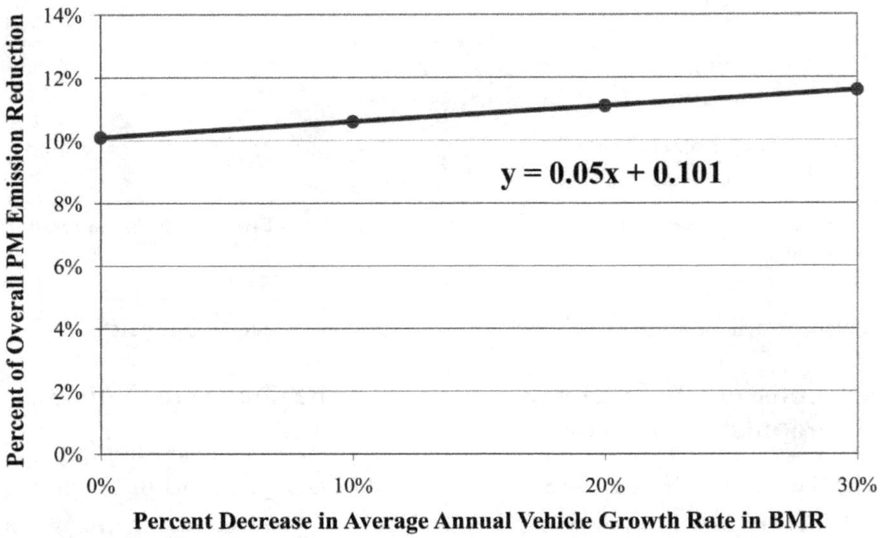

Figure 6. Impact of the change in vehicle growth rate on the overall PM_{10} emission reductions by the I/M programs.

Table 6. Variables showing the greatest contribution to variance.

Rank	Variable	Contribution to Variance
1	Problem vehicle identification rate of light-duty trucks (IdenRate)	58.4%
2	Problem vehicle as percent of total PM_{10} emission of light-duty trucks	10.0%
3	Problem vehicle identification rate of buses (IdenRate)	6.4%
4	Percent of repair work initially effective of light-duty trucks (GoodRep)	5.5%
5	Problem vehicle identification rate of motorcycles (IdenRate)	3.3%
6	Percent of excess emissions reduced by repairs of light-duty trucks (ExEm)	3.1%
7	Problem vehicle illegal operation rate of light-duty trucks (IllegalVeh)	2.4%
8	Problem vehicle identification rate of city trucks (IdenRate)	1.8%
9	Program participation rate of light-duty trucks (PartiRate)	1.8%
10	Percent of durable repairs of light-duty trucks (DurRep)	1.2%

uncertainty in this premise is large (falls into the range of 0% - 100%, see **Table** 1). Moreover, the problem vehicle identification rates of several other vehicle types (buses, city trucks and motorcycles) are also among the top 10 premises that contribute the greatest to the total uncertainty. Therefore, policy design considerations need to focus on increasing problem vehicle identification rate in order to narrow its uncertainty and improve its effectiveness.

3.6. Summary of Sensitivity Test Results

Table 7 summarizes the sensitivity test results. It indicates that the level of PM_{10} emission reductions available from the I/M programs is the most sensitive to the variable "problem vehicle identification rate (IdenRate)", since when increasing a variable from its lower limit to the upper limit while holding all the other variables constant, the greatest change happened with this variable (increased from 0.6% to 20.6%). Also, the uncertainty analysis demonstrates that the same variable "IdenRate" contributes the greatest to variance. Moreover, the variables associated with light-duty trucks play a relatively major role on the effectiveness of the I/M programs due to its role as the largest contribution to total PM_{10} emissions from motor vehicles. These findings suggest that program effectiveness can be improved by narrowing the uncertainty in the problem vehicle identification rate and by identifying a greater percentage of problem vehicles. Also, attention should be directed toward the light-duty diesel vehicle fleet in introducing the programs.

Over all, the sensitivity analysis performed here indicates that, in order to increase the problem vehicle identification rate, a key point is to improve testing procedure to maximize the ability of the programs to detect vehicles that need emission repairs. Second, studies have suggested that using more stringent testing cut-points can increase the percent of problem vehicles that are discovered by the inspection [2]. Further research on how to maximize the problem vehicle identification rates associated with I/M programs is warranted.

4. Conclusions

Since air pollution control usually imposes substantial costs on a society, an understanding of the link between specific control policies and associated health benefits would provide valuable information to decision-makers. Despite the fact that the actual effects of some mitigating policies, such as the in-use vehicle inspection and maintenance (I/M) programs, may be greatly uncertain, this uncertainty issue has traditionally been ignored in evaluating the impacts of a policy on public health. An important uncertainty issue, namely, the uncertainty about the actual effects of pollution mitigating policies, has traditionally been ignored in assessing the public health benefits of control policies. By assuming full implementation of a policy measure, an evaluation may considerably overestimate the health benefits achieved by that policy, or simply shift the focal point of decision-making processes, if the emission savings are in fact considerably

Table 7. Sensitivity test of the percent emission reductions to the key design variables.

<table>
<tr><th colspan="2">Variable</th><th>Best Estimate</th><th>Sensitivity Test Range</th><th>Minimum Performance to Achieve the 4% PM_{10} Emission Reduction Target</th><th>Sensitivity Tests Result (Change in % Emission Reductions)</th></tr>
<tr><td colspan="2">Failure cut-points</td><td>75% of the baseline emission rates</td><td>60% - 100% of the baseline emission rates</td><td>N/A</td><td>11.1% - 9.8%</td></tr>
<tr><td colspan="2">Program participation rate</td><td>90%</td><td rowspan="6">0-100%</td><td>>30%</td><td>0.6% - 11.7%</td></tr>
<tr><td colspan="2">Problem vehicle identification rate</td><td>50%</td><td>>17%</td><td>0.6% - 20.6%</td></tr>
<tr><td rowspan="3">Effectiveness of failed vehicle repairs</td><td>Percent of repair work initially effective</td><td>72%</td><td>>22%</td><td>1.1% - 14.3%</td></tr>
<tr><td>Percent of excess emissions reduced by good repairs</td><td>81%</td><td>>25%</td><td>1.1% - 12.9%</td></tr>
<tr><td>Percent of good repairs that remain durable</td><td>86.5%</td><td>>26%</td><td>1.1% - 12.1%</td></tr>
<tr><td colspan="2">Failed vehicle illegal operation rate</td><td>10% for buses and heavy trucks; 20% for light trucks and motorcycles</td><td><75%</td><td>1.1% - 12.6%</td></tr>
<tr><td colspan="2">Vehicle population growth</td><td>10% decrease in average annual growth rate under the baseline scenario</td><td>0% - 30% decrease in average annual growth rate</td><td>N/A</td><td>10.1% - 11.6%</td></tr>
</table>

uncertain for decision makers or likely to be far less than anticipated. Consequently, despite that I/M programs have been traditionally ranked superior among various vehicle emission control measures by the results of Cost-Benefit Analysis (CBA) or Cost-Effectiveness Analysis (CEA), these programs may be suboptimal if the uncertainty issue mentioned above is taken into account. Given this, this study developed a new cost-benefit analysis framework for evaluating the effectiveness of I/M programs. This framework takes into account the effects of various program design considerations, such as program participation rate, identification rate and effective repair rate, on the health benefits of policy implementation, and examines what are the minimum implementation requirements that at least ensure the benefits are greater than the costs of implementing the programs.

Applying the framework to a PM-oriented I/M program targeting all diesel-fueled vehicles in the city of Bangkok, Thailand, it was found that the health benefits achieved from the program are sensitive to several key program design elements, including participation rate and problem vehicle identification rate, fraction of effective repairs and illegal operation rate. Other variables, such as the testing cut-points and vehicle population growth rate, only have modest effects on the overall emission reduction and consequent health benefits. Overall, the performance of multiple variables associated with I/M program design needs to be improved simultaneous in order to achieve the targeted benefits of the program.

The main limitation of using the new analysis framework to evaluate the effectiveness of the I/M programs is that presently there is very limited information on the performance of the important program design elements globally. The findings from running the "I/M Design" spreadsheet could be improved when more empirical data worldwide for the input variables are collected. Secondly, one source of emission reductions achieved by the I/M programs is improved maintenance of vehicles in anticipation of the required inspection process, but this kind of emission reduction relative to the baseline is not considered in estimating the emission reductions delivered by the programs. This portion of emission reductions can be large for diesel-fueled vehicles since pollution levels from these vehicles are heavily dependent on maintenance, perhaps resulting in an underestimate of the emission reduction benefits of the programs. Given the information gap, further research is warranted to examine the potential emission reductions due to improved maintenance by vehicle owners in anticipation of required I/M testing. Finally, further research needs to concentrate on the question that how to design and implement a PM-oriented I/M program in order to improve the performance of the key variables found in this study, e.g. the problem vehicle identification rate associated with an I/M program, so that more concrete and practical advice is provided to decision makers based on the theoretical conclusions of the present study.

References

[1] Posada, F., Yang, Z. and Muncrief, R. (2015) Review of Current Practices and New Developments in Heavy-Duty Vehicle Inspection and Maintenance Programs. The International Council on Clean Transportation White Paper, Washington DC.

[2] Eisinger, D.S. (2005) Evaluating Inspection and Maintenance Programs: A Policy-Making Framework. *Journal of the Air & Waste Management Association*, **55**,

147-162. https://doi.org/10.1080/10473289.2005.10464609

[3] National Research Council (2001) Evaluating Vehicle Emissions Inspection and Maintenance Programs. National Academies Press, New York.

[4] Harrington, W. (1997) Fuel Economy and Motor Vehicle Emissions. *Journal of Environmental Economics and Management*, **33**, 240-252. https://doi.org/10.1006/jeem.1997.0994

[5] Beaton, S.P., Bishop, G.A., Zhang, Y., Ashbaugh, L.L., Lawson, D.R. and Stedmam, D.H. (1995) On-Road Vehicle Emissions: Regulations, Costs, and Benefits. *Science*, **268**, 991-993. https://doi.org/10.1126/science.268.5213.991

[6] Sterner, T. (2003) Policy Instruments for Environmental and Natural Resource Management. Resource for the Future Press, Washington DC.

[7] World Bank (2001) Vehicular Air Pollution: Setting Priorities, South Asia Urban Air Quality Management Briefing Note No. 1. http://siteresources.worldbank.org/PAKISTANEXTN/Resources/UrbanAir/VehicularAirPollution.pdf

[8] PA Government Services, Inc. (2004) Vehicle Inspection and Maintenance Programs: International Experience and Best Practices. Report, The US Agency for International Development Office of Energy and Information Technology, Washington DC.

[9] Li, Y. and Crawford-Brown, D.J. (2011) Assessing the Co-Benefits of Greenhouse Gas Reduction: Health Benefits of Particulate Matter Related Inspection and Maintenance Programs in Bangkok, Thailand. *Science of the Total Environment*, **409**, 1774-1785. https://doi.org/10.1016/j.scitotenv.2011.01.051

[10] World Bank (2003) Thailand: Reducing Emissions from Motorcycles in Bangkok. Report 275/03, World Bank Energy Sector Management Assistance Programme.

[11] USEPA, United States Environmental Protection Agency (2007) Cleaner Diesels: Low Cost Ways to Reduce Emissions from Construction Equipment. United States Environmental Protection Agency National Center for Environmental Innovation.

[12] Crawford-Brown, D.J. (1999) Risk-Based Environmental Decisions: Methods and Culture. Kluwer Academic Publishers, Boston, MA. https://doi.org/10.1007/978-1-4615-5227-7

7

Biological Soil Quality Indicators and Conditioners in a Plant-Assisted Remediation of Crude Oil Polluted Farmland

E. O. Nwaichi[1*], L. C. Chuku[2]

[1]University of Massachusetts, Amherst, MA, USA
[2]University of Port Harcourt, Port Harcourt, Nigeria
Email: *nodullm@yahoo.com, lcchuku@gmail.com

Abstract

Owing to vital roles played by enzymes in the preservation of the make-up of soil ecosystem and functional diversity, the influence of organic manure on the resultant biological quality of a crude oil polluted agricultural soil from a 90 d phytoremediation pot experiment was investigated. A 4-factor phyto-assisted clean-up of crude oil polluted agricultural soil was designed with options of manure amendments, to boost micobial activities. Profiles of *β*-glucosidases, proteases, dehydrogenases, phosphomonoesterases and respiration were investigated. Analysis of variance of triplicate experiment was carried out. Application of soil conditioner gave no marked dehydrogenase activity, which increased with depletion of available phosphorus. Marked increases in CO_2 release and alkaline phosphatase activity with soil conditioning may implicate beneficial relationship with the abundance of microbial populations. Flooding in some pots correlated with *β*-glucosidase and respiratory acitivities. A direct relationship between cellulose breakdown, measurable with *β*-glucosidase activity, organic matter and CO_2 release, measurable with respiratory activity within all soils was found in the present study. Use of organic manure significantly improved CO_2 release by soil biota in hydrocarbon-impacted soil and may be explored for phytoremediation technique.

Keywords

Soil Quality, Enzymes and Activities, Environmental Changes, Organic Matter, Organic Manure, Petroleum Stress

1. Introduction

Microorganisms and enzymes basically drive mineralization, degradation of organic matter and nutrient recycling, and collectively constitute soil biological processes. The possibilty of understanding changes in soil organic matter using enzyme activities is well described by Gianfreda and Ruggiero [1] and may be useful in the face of increased urbanization and industrialization with accompanying environmental hazards. Enzyme biochemical assays do not show actual lotus extent of activity but represent potential activity and hence have been described by Bandick and Dick [2] as indicators. This may have stemmed from their usual indirect measurements by determining their activity using biochemical assays in the laboratory. β-glucosidase, an enzyme involved in cellulose degradation, has been reported to be quick to respond to soil management [3] [4] and linked to its involvement in soil organic carbon cycle and thus gives an early sign of yield and changes associated with organic matter. Piotrowska and Koper [5] underscored the significance of the product of enzymatic hydrolysis of cellulose, being the most bountiful organic compound in the biosphere, as a source of energy for soil microorganisms. Protein degradative pathway or proteolysis starts with the hydrolysis of peptide bonds linking amino acids in the protein, forming polypeptide chain. This process, occurring in viruses, bacteria, archea, plants and animals, is catalysed by the enzyme called protease often named peptidase or proteinase [6]. They are involved in the invertebrate prophenoloxidase-activating cascade. Intracellular enzymes and dehydrogenases however oxidize substrates by relocating hydride(s) (H^-) to an oxidising agent, normally a flavin coenzyme e.g. FAD or FMN or NAD^+/$NADP^+$. Enzyme activity is measured as a mounting effect of continuing microbial activity and agitation of the tenable population at sampling. Dick [7] presented dehydrogenase as a typical illustration of an enzyme depicting activity of viable cells, which in theory cannot take place in settled soil complexes. Dehydrogenases activity (DHA) is an indicator of the microbiological redox-systems and provides important determination in the soil samples, and gives useful information concerning biological properties of soil. Dehydrogenases are chiefly produced by anaerobic microorganisms but can utilize oxygen and other electron acceptors. Wolińska and Stępniewska [8] reported that under anaerobic conditions, activities of soil DHA increase greatly. They reiterated the possible influences by soil pH, depth, temperature, organic matter content, available oxygen, oxidation-reduction potential, heavy metal pollution, season of the year, pesticide use, etc. Phosphatases describe a group of enzymes driving the hydrolysis of esters and anhydrides of orthophosphoric acid, thus performing a salient role of transforming organic P in litter into inorganic phosphate (HPO_4^{2-}, $H_2PO_4^{4-}$) in soil, and this is readily available to soil and plants organisms. They have been reported to derive from plant roots and soil microorganisms with varying kinetic properties [9], and differ in their response to anthropogenic influences. They are also, predictors of soil function (nutrient cycling) with plant available phosphorus as significance. Absence of, reduced or poor enzyme activity may lead to an accumulation of dangerous chemicals in the environment that could hamper the activity of soil enzyme [7] [10]. These activities can better describe the response of phosphorus transformation to climatic and other environmental influences. Respiration is the release of carbon via heterotrophic consumption from the soil in the form of CO_2 and is a major ecosystem process. Nwaichi *et al.* [11] identified the dire need for provision of soil mechanism to solve the interwoven sustainable development and environmental crises in the Niger Delta, given high oil exploration and exploitation activities in the region. Limited accounts on the quality of crude oil polluted soils exist, to our knowledge as well as quality of wholly or partially recovered soils. This study therefore seeks to evaluate possible effect of management practices on vital and relevant enzyme activities in petroleum polluted soil with a four-factor-phytoremediation recovery attempts.

2. Materials and Methods

2.1. Experimental Design

Soil samples were taken from a freshly (2 weeks) and burnt crude oil spilled farmland in Oshie community of Akalaolu (a major host community to a big oil company in Nigeria) Ahoada West in the Niger Delta, after a field survey. Cause of spill was traced to sabotage. This experiment was designed to follow the 2013 planting season in Nigeria in a four-factor phytorediation experiment and soil depth of 20 cm was chosen to have a beter community representation. The design was a 90-d phytoremediation triplicate pot experiments using *V. subterranea, H. brassilensis, C. citratus* and *F. littoralis*. Experimental control consisted of an agricultural soil in the same region with history of no pollution and poultry dung was used as soil amendment. Need-based watering was adopted and thinned plants were left on the soil on which it was grown. The amount of conditioner (animal dung) in this design was determined following Nwaichi *et al.* [12]. After 90 ds, collected rhizopheric soils were shipped in ice chest coolers to Environmental Microbiology laboratory at the Institute of Agrophysics lublin Poland for analyses and were subjected to protocols outlined below:

2.2. β-Glucosidase Activity

The activity of β-glucosidase was measured as given by Eivazi and Tabatabai [13]. This method is based on the spectrophotometrical measure of released p-nitrophenol, PNP after 1 hour p-nitrophenyl-β-D-glucopyranoside (PNG)—laden soil incubation at 37°C with universal buffer as the substrate, though with modifications (pH 6.0). Measured enzyme activity was calculated as moles PNP released per kg dry soil per 1 hour (mM pNPkg^{-1}h^{-1}). Chemical properties, like organic carbon content (C_{ORG}), availabe P, total nitrogen content (N_{TOT}) and pH in 1 mol KCl dm^3, were determined according to standard methods accepted in soil science.

2.3. Protease (Adapted from Reported Protocol by Alef and Nannipieri [14] and Tabatabai [15]

Using a weighing balance (RADWIG PS1200/C/2 Max 1200 g Min 0.5 g e = 0.1

g, d = 0.01 g), 2 g of seived (2 mm) moist soil were added to 15 ml Falcon (centrifuge) tubes (3 replicates and 1 control). Five millilitres of 1% sodium caseinate in TRIS HCl buffer pH 8.1 substrate (prepared over night and refrigerated), was added into test only to measure the amino acids released after soil incubation with sodium caseinate for 2 hr at 50°C using Folin-Ciocalteu reagent. Only TRIS HCl buffer pH 8.1 (5 ml) was introduced into control. This was shaken for 1 hr at 50°C and tubes placed in cold water to cool. An volume of 2 ml 17.5% TCA was introduced to test and control followed by centrifugation @ 3000 rpm for 2 mins. Into 2 ml supernatant in 5 ml test tube, 3 ml 1.4 M Na_2CO_3 was introduced to all tubes and was shaken thoroughly, then 1 ml Folin-Ciocalteu reagent (diluted 3 times) was added. After centrifugation for 2 mins at 200 rpm, absorbance readings (578 nm) of 96 Corning plates containing 200 μl samples in Infinite M200 PRO TECAN Spectrophotometer were taken. Buffer was used as blank. Calibrattion curve was prepared, measured absorbance for the controls corrected, then calculation was done as follows: protease activity (mg tyrosine kg $d{\cdot}m^{-1}h^{-1}$) = (C × 15)/dwt, where dm is dry weight of 1 g of soil, 15 is the final volume of solutions introduced to the soil during assay and C is the measured tyrosine concentration.

2.4. Dehydrogenases Activity (DHA)

DHA was measured following Casida *et al.* [16] where specific dye, triphenyltetrazolium chloride (TTC) that can specify the flow of electrons was employed as indicator of electron transport system (ETS) activity. Colorless and water soluble substrate (TTC) was reduced by dehydrogenases present in the soil environment, and this generates a red-colored insoluble product, triphenylformazan-TPF that was quantified colorimetrically within the range of visible light (485 nm). TRIS-HCl pH 7.4 was used instead of $CaCO_3$ for buffering soil system. Developed colors (red) of high intensity correlated with higher activity in soil samples prepared for spectrophotometry while no-red coloration or light colors (red) pe-indicated lower DHA values.

2.5. Phosphomonoesterases Activity

Determination of p-nitrophenol released after soil incubation with p-nitrophenyl phosphate for 1 hr at 37°C is the underlying principle. Two sets of 1 g air-dried soils in Falcon tubes (3 replicates and 1 control) were prepared using weighing balance for alkaline and acid phosphatases. An aliquot of 0.25 ml Toluene was added to 1 g dry test and control soil and in a fume chamber 10 mins later, 4 ml buffer was introduced. To test, 1 ml PNP substrate was added and all samples were incubated at 37°C for 1 hr. Following this, 1 ml PNP was added to control. Again, an aliquot of 1 ml 0.5 M $CaCl_2$ and 4 ml 0.5 M NaOH were introduced to all samples and shaken in a Multi RS-60 BIOSAN Programmable rotator-mixer/shaker for 3 mins before Ependorf Centrifuge 5810 R centrifugation at 4000 rpm for 10 mins. At 485 nm using buffer as blank, microplates containing

200 μl samples were read off in a spectrophotometer. Following preparation of calibration curve, control results were corrected and the PNP per ml of the filtrate were calculated by making reference to the calibration curve as in this formula:

$$\rho - \text{Nitrphenol}\left(\text{mmol PNP kgd} \cdot \text{m}^{-1}\text{h}^{-1}\right) = \left(C \times v\right) / \left(dwt \times SW \times t\right)$$

C = concentration of ρ-nitrophenol μg ml^{-1} filtrate measured, dwt = dry weight of 1 g moist soil, v = total volume of soil suspension in ml, SW = soil weight, t = incubation time in hours.

For buffer preparation, 0.1 M HCl and 1 M NaOH mixture were corrected to pH 6.5 and 11 for acid and alkaline phosphatases respectively.

For substrate preparation, 1 ml acid buffer pH 6.5 was added to 0.05 g PNP (in a shielded beaker) per sample for acid phosphatase assay and 1 ml acid buffer pH 11 to 0.05 g PNP for alkaline phosphatases.

2.6. Respiratory Activity

Substrate (Glucose) induced method adapted from reported protocol by Alef and Nannipieri [14] was followed. To respiratory flasks in triplicate, 10 g < 2 mm screen soil was added to the outer jar and to the inner jar, 2.5 ml 0.2 M NaOH was added. One millilitre of glucose solution was evenly dropped onto the soil. Only 10 g soil-containing flasks for control, were autoclaved using Fedegari Autoclave AG and Classic Prestige Medical Autoclave for 20mins at 121°C. Thereafter, 2.5 ml 0.2 M NaOH and 1 ml glucose solution were introduced. These were left to stand for 24 hr at room temperature. In a UV-lit gas chamber, 1 ml 1 M $BaCl_2$ was added to to a 25 ml beaker containing transferred NaOH. Using 0.1 MHCl and Phenolphthalein indicator, titration was carried out while swirling and for calculation titre values were noted.

2.7. Statistical Analysis

Means of triplicate data were analysed using Analysis of Variance, ANOVA. STATISTICA v 10 were employed to run Factor and Principal component analyses.

3. Results and Discussion

Organic fertilization is known to beneficially increase organic C and N concentration in soil and affects organic matter in terms of quality and quantity. The pH values measured in 1 mol KCl dm^3 ranged from 3.45 (in initial polluted soil) - 6.8 (after 90 ds soil from manured Bambara cultivated regime). Total nitrogen content reached 1.7% in soil cultivated with Bambara and amended.

The class (Leguminosae) may have placed Bambara cultivated soils at an advantage in this regard. Observed influence of organic manure amendment in β-Glucosidase activity, BGA is as shown in **Figure 1**. Rubber-cultivated soils, when compared among plants, were significantly higher. Activity among plant-cultivation types and treatments responded positively to soil conditioners

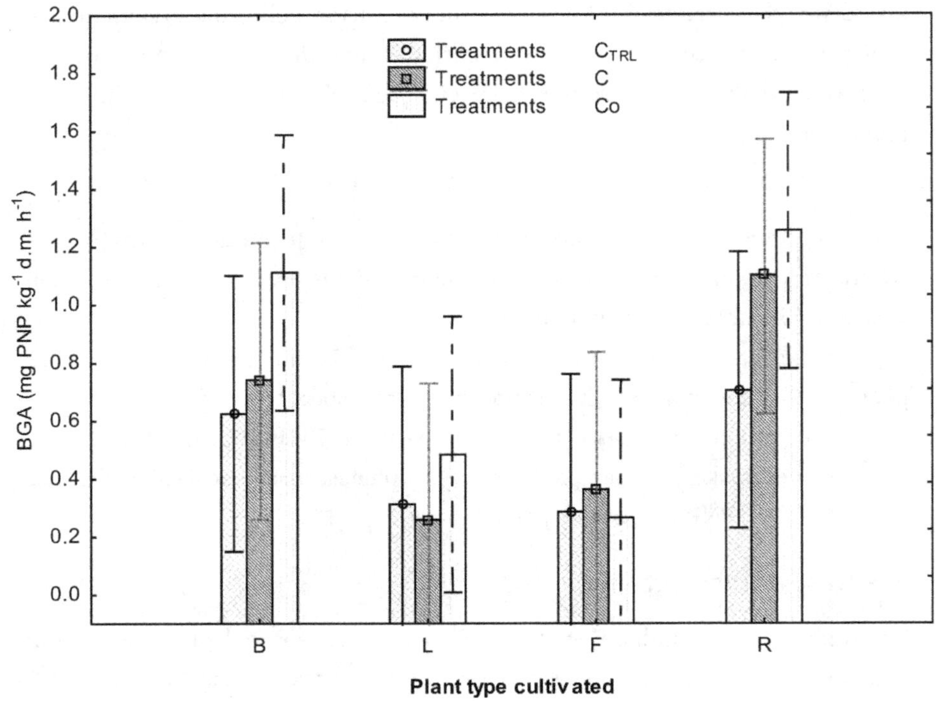

Figure 1. Effect of Organic manure application on *β*-glucosidase activity, BGA in different study soils. Vertical bars denote 0.95 confidence intervals. Values are means (n = 3) ± SD. B, L, R and F denotes Bambara, Lemon grass, Rubber, and Fimbristylis- cultivated soils; attached ctrl, c, and co denotes control (unpolluted), crude oil polluted and organically amended-crude oil polluted agricultural soils respectively. Increased activity with conditioner but for F-cultivated soil.

for Bambara, Lemon grass and Rubber-cultivated regimes but were not marked. This validates the environmentally friendliness of phytoremediation as a technique of soil clean-up. There was flooding in pots containing Fimbriatylis-cultivated designs and may have contributed to observed correlation with BGA activity. Protease activity (**Figure 2**) were highest in unpolluted soils for all plants.

Application of soil condition to phytoremediation design improved protease activity, though not marked ($p < 0.05$) and is indicative of community response to Carbon, Nitrogen and Sulphur limitations [17]. Observed rate of consumption for TTC-specific substrate for dehydrogenases, DHA, did not decrease for unamended contaminated designs leading to lower DHA activity (**Figure 3**).

It is noteworthy that these regimes were more acidic than control (unpolluted) and organic manure-amended counterparts. Levyk *et al.* [18] demonstrated in their work that acidic soil conditions in the pH range of 1.5 - 4.5 resulted in strong DHA inhibition. Weakening of enzymatic activity with increased soil acidity was linked by Frankenberger and Johanson [19] with the subsequent destruction of hydrogen bonds and ion in corresponding enzyme active centre. The increases with soil conditioner were however, not marked. Heavy metals such as Cd, Pb and As (data not shown) were observed at initial characterization to exceed their set limits in agricultural soils. Their presence may have reduced DHA as they can interact with the enzyme-substrate complex,

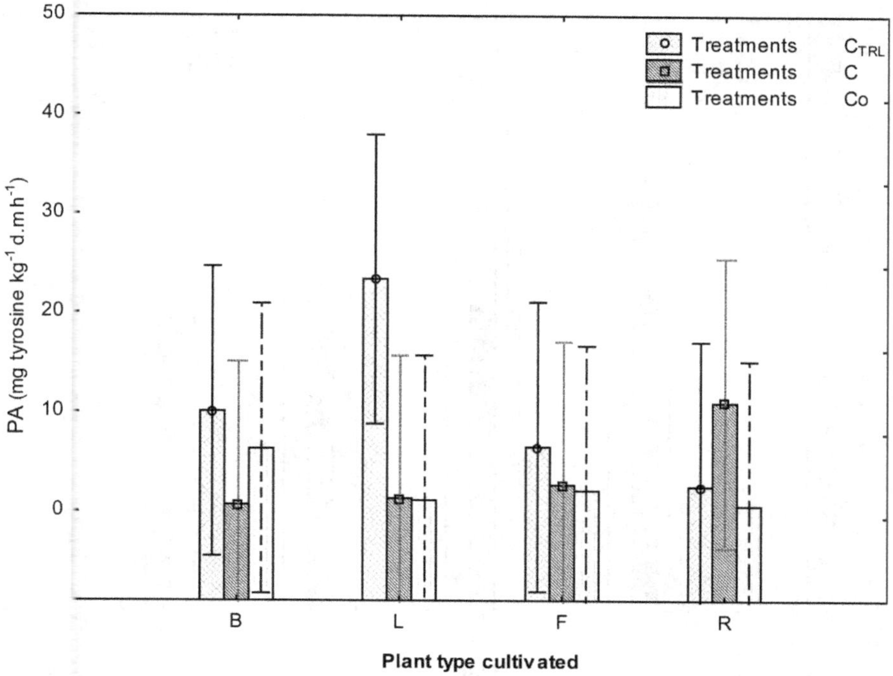

Figure 2. Protease activity, PA pattern in studied variants. PA denotes protease activity. No significant influences with conditioner was observed.

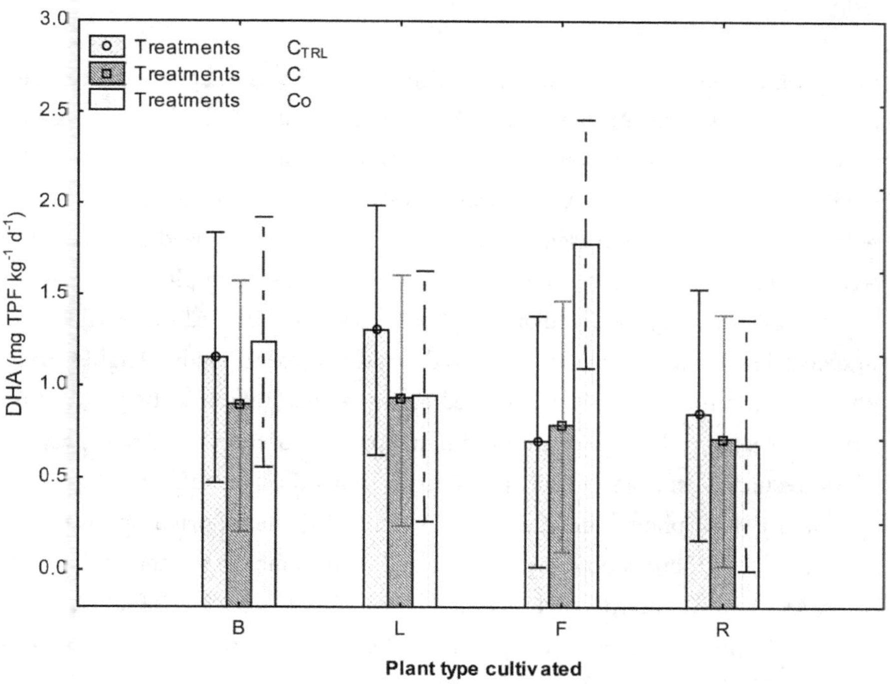

Figure 3. Dehydrogenases activity, DHA among plants and treatments. Vertical bars denote 0.95 confidence intervals. No significant influences with conditioner.

bringing about denaturation of the enzyme protein or interactions with the protein-active groups. We therefore think that DHA and microbial biomass may not be proportional given the peculiarity and complexity of contaminated soil with varied activity.

The highest acidic phosphatase activity, ACPA (**Figure 4**) was measured in

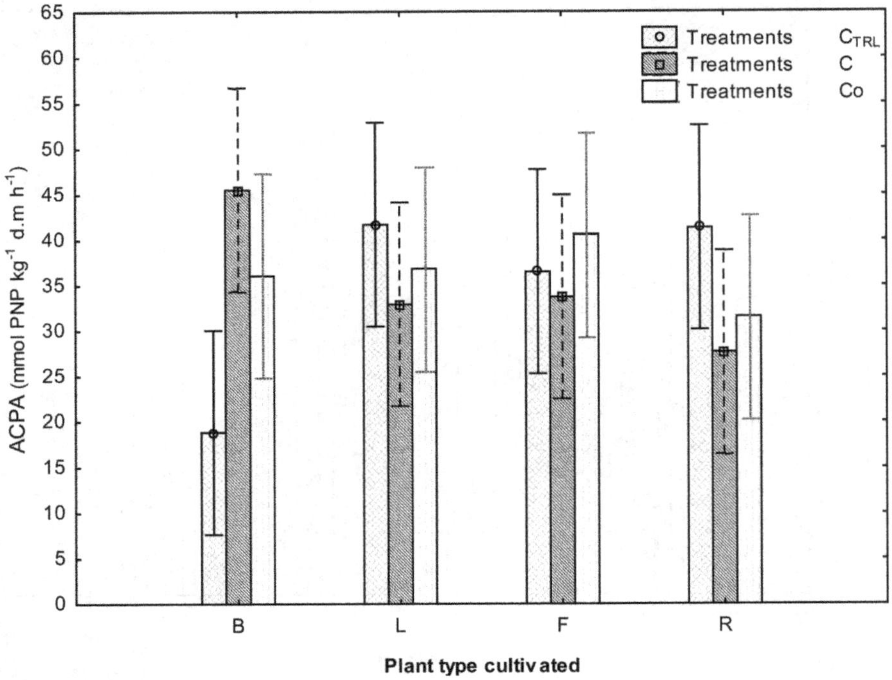

Figure 4. Acid phosphatase activity, ACPA among plants and between treatments as influenced by conditioner. Influence of conditioner was significant only in B-cultivated soils.

the unamended Bambara cultivated soils and was influenced by depleted available P (**Figure 9**) measured. Nannipieri *et al.* [20] showed negative correlations between this enzyme activity and available P in sandy soils. Our study soil is a clayey loam. A significant connection was also observed in activity with organic matter-increased organic matter content improved ACPA. There were no significant increases observed with soil conditioner application.

As shown in **Figure 5**, alkaline phosphatase activity, ALPA was low or undetectable in most soils investigated in the present study. Higher levels obtained for amended Bambara planted regime is attributable to its pH_{KCl} of 6.8 (data not shown), which exceeded maximum optimum of 6.5 for acid phosphatases. These results do not agree with previous findings that soil pH correlates with optimum pH for phosphomonoesterase activity [21]. As reported by Eivazi and Tabatabai [13], there is no excretion of alkaline phosphomonoesterases by plant roots, so we attribute activity to soil microorganisms. No clear influence of the clean-up technique was evident on soil ALPA. However, when organic manure was applied, significant differences were found.

Similarly, comparing respiratory activity, RESA among plant type cultivated (**Figure 6**), coincided with observed BGA pattern. The choice of Bambara and native specie, Rubber ere most beneficial. We found a direct relationship between cellulose breakdown, measurable with BGA, organic matter content and CO_2 release, measurable with RESA within all soils in the present study. These observed marked increases (**Figure 7**) with soil conditioning may implicate beneficial relationship with the abundance of microbial populations.

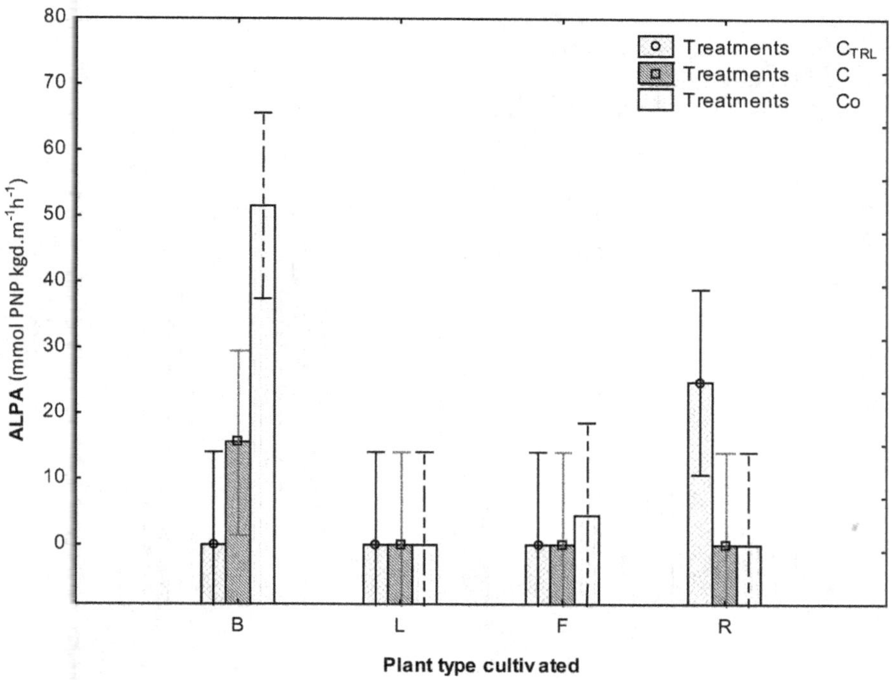

Figure 5. Alkaline phosphatase activity, ALPA among plants and between treatments as influenced by conditioner.

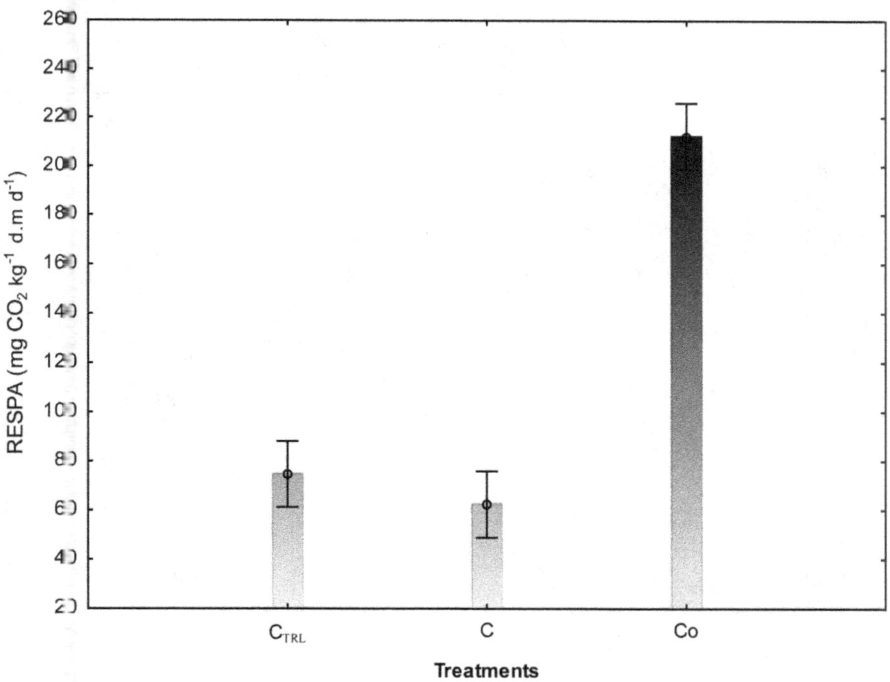

Figure 6. Respiratory activity, RESPA as influenced by soil conditioner among plant types cultivated.

Soil respiration, also known as carbon mineralization, was statistically marked among treatments and cultivated plant types (**Figure 6**) and gives an indication of the innate ability to support soil life including soil microorganisms, animals and crops by study soil, espectially given application of conditioners. There was no activity prior to phytoremediation experiment. Also, this life-support

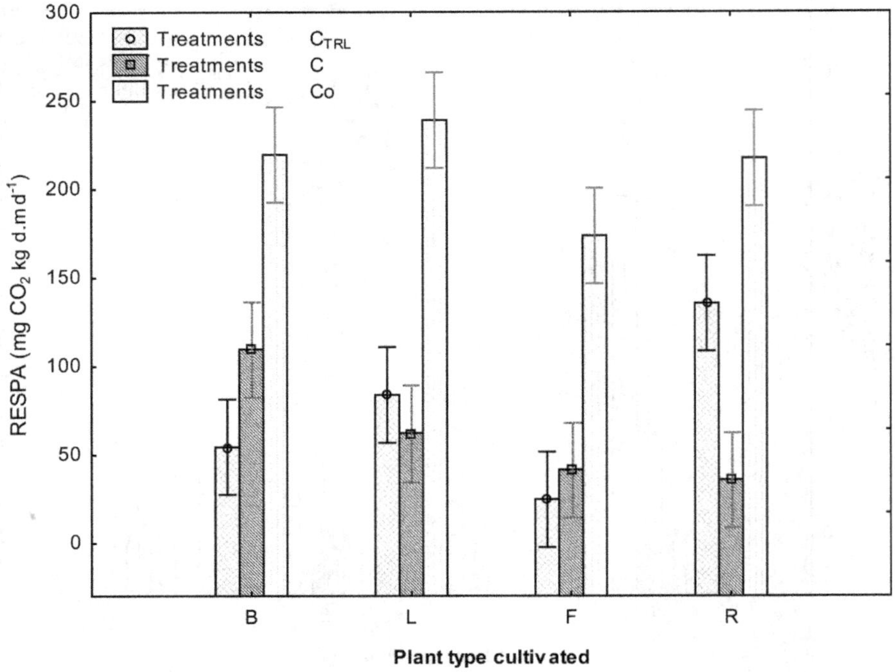

Figure 7. Respiratory activity, RESPA as influenced by soil conditioner among plants and treatments.

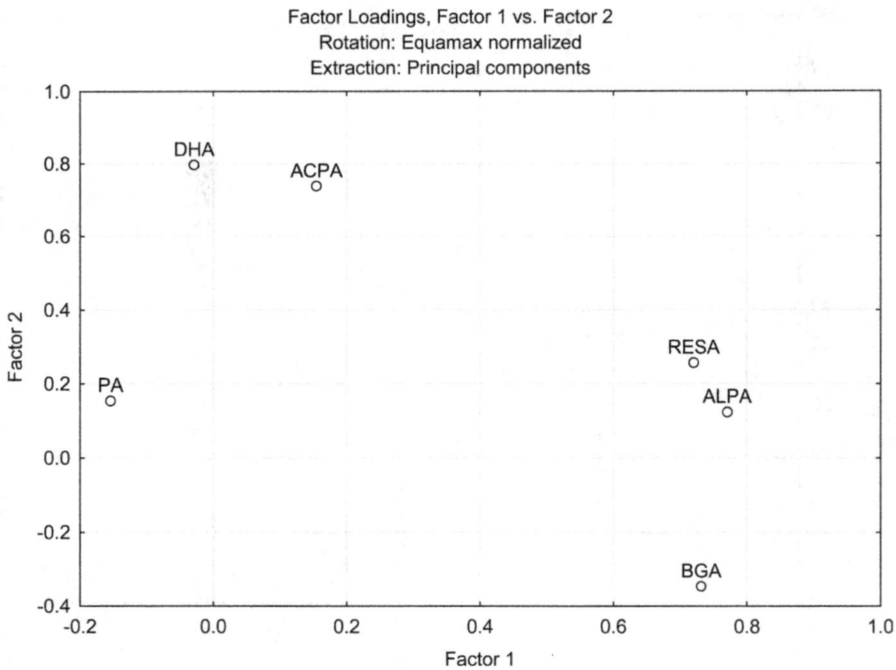

Figure 8. Multivariate factor analysis of expressed activities.

capacity showed no positive effect owiththe application of organic manure for Fimbristylis-cultivated soils. Frequent flooding of pots cultivated with Fimbristylis may have induced incomplete mineralization of soil organic matter and resultant decreased aerobic soil microbial activity.

Factor analysis (**Figure 8**) was used to identify "invisible" factors, that portray the hidden "organizing principle" of anything being measured with a number of observable counts like dehydrogenases, proteases, phosphomonoesterases, β-glucosidases and respiratory activities. Factor scores show how each "hidden" factor is analogous with the "observable" variables considered in this analysis. Factor loadings of 0.77, 0.72 and 0.73 show that alkaline phosphatases, respiratory and β-glucosidases activities can be used to give detailed account of hidden Factor 1; in other words, Factor 1 has features very similar to aforementioned activities. Other observable counts were not helpful in describing Factor 1. Similarly, Factor loadings of 0.80 and 0.74 show that Factor 2 has features very similar to dehydrogenases and acid phosphatases activities.

For organic matter (**Figure 9**), significant increases was observed with conditioner but for F-cultivated soil but were not marked (**Figure 10**) with

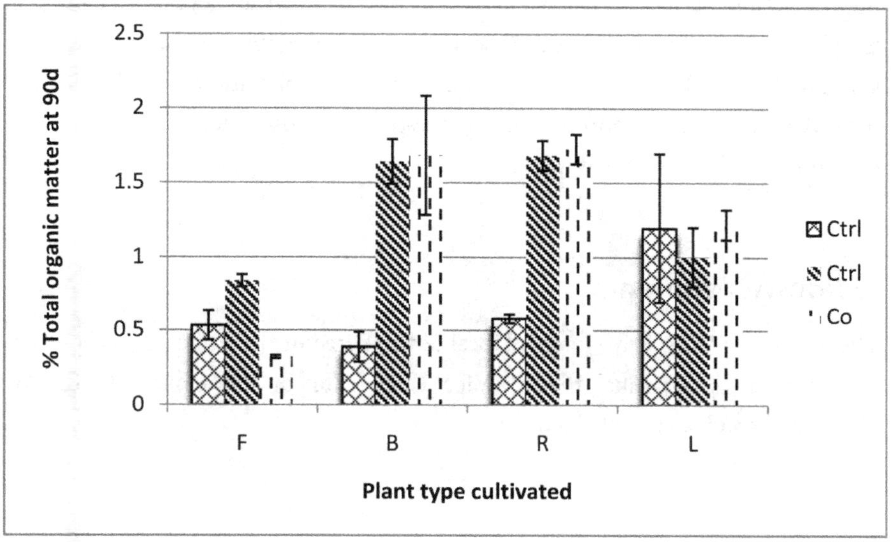

Figure 9. Total organic matter, TOM at 90 d of phytoremediation. Significant increases was observed with conditioner but for F-cultivated soil.

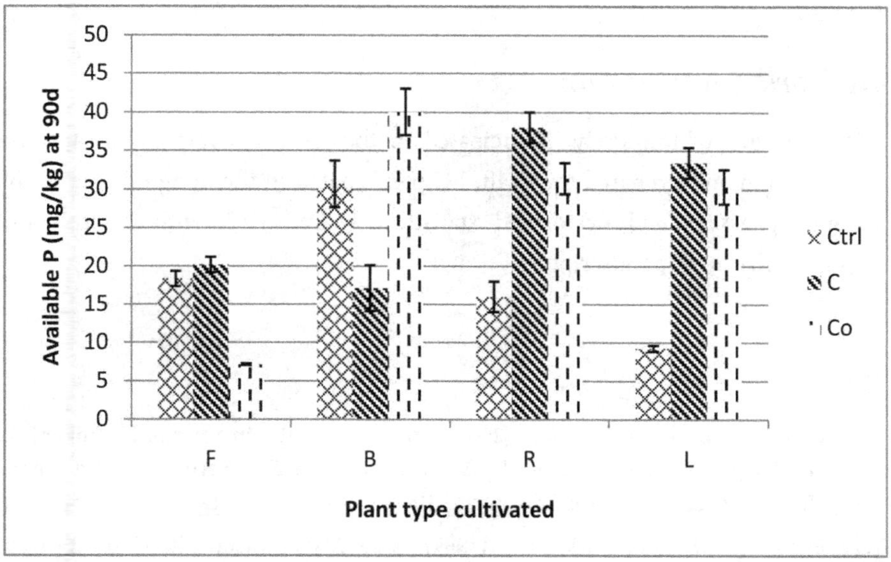

Figure 10. Available Phosphorus at 90 d of phytoremediation. Non-marked increases with conditioner except for F-cultivated soils.

observed levels of available Phosphorus except for F-cultivated soils.

Generally, results indicate the importance of incorporating leguminous plants in phytoremediation site and the influence of plant-soil-microbial interactions where there is a nutrient boost.

4. Conclusion

Reduced activity of proteases may be due to autolysis with time lag between sampling and analysis in different countries and may constitute limitation to this study. Higher β-glucosidase activity was observed to be connected with organic matter content and respiratory activity. The legumes may be useful in phytoremediation design given observed trend with Bambara plants. Application of organic manure significantly improved CO_2 release by soil biota in hydrocarbon-stressed soil and may be explored for phytoremediation technique. Measurement of inherent changes in biological soil properties, therefore could complement microbial indicators for a more sensitive and rapid approach in understudying soil quality, especially the environmentally impacted types.

Acknowledgements

The assistance from Prof. Joseph Horabik the Director, Institute of Agrophysics, and Prof. J. A. Ajienka, the 7th Vice Chancellor of the University of Port Harcourt, is well acknowledged.

Funding

The authors are grateful to UNESCO LO'real FWIS for financial support.

Authors' Contribution

EO conceived of the study, participated in the design, laboratory assays and coordination and prepared the draft. LC participated in the design of the study, laboratory assays and performed statistical analysis. All authors read and approved the final manuscript.

References

[1] Gianfreda, L. and Ruggiero, P. (2006) Enzyme Activities in Soil. In: Nannipieri, P. and Smalla, K., Eds., *Nucleic Acids and Proteins in Soil*, Springer-Verlag, Berlin Heidelberg, 257-310. https://doi.org/10.1007/3-540-29449-X_12

[2] Bandick, A.K. and Dick, R.P. (1999) Field Management Effects on Enzyme Activities. *Soil Biology and Biochemistry*, **31**, 1471-1479. https://doi.org/10.1016/S0038-0717(99)00051-6

[3] Stege, P.W., Messina, G.A., Bianchi, G., Olsina, R.A. and Raba, J. (2010) Determination of Beta-Glucosidase Activity in Soils with a Bioanalytical Sensor Modified with Multiwalled Carbon Nanotubes. *Analytical and Bioanalytical Chemistry,* **397**, 1347-1353. https://doi.org/10.1007/s00216-010-3634-7

[4] Russel, S., Górska, E.B. and Wyczó£Kowski, A.I. (2005) The Significance of Studies on Enzymes in Soil Environment. *Acta Agrophysica,* **3**, 27-36.

[5] Piotrowska, A. and Koper, J. (2010) Soil *β*-Glucosidase Activity under Winter Wheat Cultivated in Crop Rotation Systems Depleting and Enriching the Soil in Organic Matter. *Journal of Elementology,* **15**, 593-600. https://doi.org/10.5601/jelem.2010.15.3.593-600

[6] van der Hoorn, R.A. (2008) Plant Proteases: From Phenotypes to Molecular Mechanisms. *Annual Review of Plant Biology,* **59**, 191-223. https://doi.org/10.1146/annurev.arplant.59.032607.092835

[7] Dick, R.P. (1994) Soil Enzyme Activity as an Indicator of Soil Quality. In: Doran, J.W., *et al.*, Eds., *Defining Soil Quality for a Sustainable Environment,* Madison, WI, 107-124.

[8] Wolińska, A. and Stępniewska, Z. (2012) Dehydrogenase Activity in the Soil Environment. In: "*Dehydrogenases*". *Biochemistry, Genetics and Molecular Biology,* Canuto, RA. https://doi.org/10.5772/48294

[9] Rejsek, K., Vranova, V., Pavelka, M. and Formanek, P. (2012) Acid Phosphomonoesterase (E.C. 3.1.3.2) Location in Soil. *Journal of Plant Nutrition and Soil Science,* **175**, 196-211. https://doi.org/10.1002/jpln.201000139

[10] Nwaichi, E.O. and Essien, E.B. (2016) Biochemical Indices in Ecology: A Catalyst in Environmental Study. *Biochemistry & Analytical Biochemistry,* **5**, 238.

[11] Nwaichi, E.O., Frac, M., Peters, D.E. and Akpomiemie, B.O. (2014) Conditioners and Significance in t-RFLP Profile of the Assemblage of Prokaryotic Microorganisms in a Niger Deltaic Crude Oil Polluted Soil. *African Journal of Biotechnology,* **13**, 4220-4225. https://doi.org/10.5897/AJB2014.13886

[12] Nwaichi, E.O., Osuji, L.C. and Onyeike, E.N. (2011) Evaluation and Decontamination of Crude Oil-Polluted Soils Using Centrosema Pubescen Benth and Amendment-Support Options. *International Journal of Phytoremediation,* **13**, 373-382. https://doi.org/10.1080/15226514.2010.495144

[13] Eivazi, F. and Tabatabai, M.A. (1988) Glucosidases and Galactosidases in Soils. *Soil Biology and Biochemistry,* **20**, 601-606. https://doi.org/10.1016/0038-0717(88)90141-1

[14] Alef, K. and Nannipieri, P. (1995) *Β*-Glucosidase Activity. In: *Methods in Applied Soil Microbiology and Biochemistry,* Academic Press, Harcourt Brace & Coy Publishers, London, 350-352.

[15] Tabatabai, M.A. (1994) Soil Enzymes. In: Weaver, R.W., *et al.*, Eds., *Methods of Soil Analysis, Part* 2: *Microbiological and Biochemical Properties,* 775-833.

[16] Casida, L., Klein, D. and Santoro, T. (1964) Soil Dehydrogenase Activity. *Soil Science,* **98**, 371-376. https://doi.org/10.1097/00010694-196412000-00004

[17] Sims, G.K. and Wander, M.M. (2002) Proteolytic Activity under Nitrogen or Sulfur Limitation. *Applied Soil Ecology,* **568**, 1-5. https://doi.org/10.1016/S0929-1393(01)00192-5

[18] Levyk, V., Maryskevych, O., Brzezinska, M. and Wlodarczyk, T. (2007) Dehydrogenase Activities of Technogenic Soils of Former Sulphur Mines (Yvaoriva Nemyriv, Ukraine. *International Agrophysics,* **21**, 255-260.

[18] Levyk, V., Maryskevych, O., Brzezinska, M. and Wlodarczyk, T. (2007) Dehydrogenase Activities of Technogenic Soils of Former Sulphur Mines (Yvaoriva Nemyriv, Ukraine. *International Agrophysics*, **21**, 255-260.

[19] Frankenberger, W.T. and Johanson, J.B. (1982) Influence of Crude Oil and Refined Petroleum Products on Soil Dehydrogenate Activity. *Journal of Environmental Quality*, **11**, 602-607. https://doi.org/10.2134/jeq1982.00472425001100040010x

[20] Nannipieri, P., Muccini, L. and Ciardi, C. (1983) Microbial Biomass and Enzyme Activities: Production and Persistence. *Soil Biology and Biochemistry*, **15**, 679-685. https://doi.org/10.1016/0038-0717(83)90032-9

[21] Badalucco, L., Grego, S., Dell'Orco, S. and Nannipieri, P. (1992) Effect of Liming on Some Chemical, Biochemical, and Microbiological Properties of Acid Soils under Spruce (*Pieca abies* L.). *Biology and Fertility of Soils*, **14**, 76-83. https://doi.org/10.1007/BF00336254

8

Optimization of Biogas Production from Organic Municipal Waste: Development of Activated Sludge as Digesters Inoculum

Mahamadi Nikièma[1*], Marius K. Somda[1], Kifouli Adéoti[2], Désiré Traoré[1], Farid Baba-Moussa[2], Fatiou Toukourou[2], Dayéri Dianou[3], Alfred S. Traoré[1]

[1]Centre de Recherche en Sciences Biologiques Alimentaires et Nutritionnelles (CRSBAN), Département de Biochimie-Microbiologie, Université Ouaga I, Pr Joseph KI-ZERBO, 03 BP 7131Ouagadougou 03, Burkina Faso

[2]Laboratoire de Microbiologie et Technologies Alimentaires (LAMITA), Département de Biologie végétale, Université d'Abomey-Calavi, B.P. 526 Abomey-Calavi, Bénin

[3]Centre National de la Recherche Scientifique et Technologique (CNRST), 03 BP 7192 Ouagadougou 03 Burkina Faso, Ouagadougou, Burkina Faso

Email: *mahamadinikiema87@gmail.com

Abstract

This study is a contribution to the optimization of organic fraction of municipal waste bioconversion into biomethane, by activated sludge production as inoculum for digesters. The wastewater (WW) and cow dung (CD) samples were taken from the slaughterhouse of Ouagadougou town, Burkina Faso. Different mixtures were made, enriched with mineral solution and cellulose at 5% (w/v) as: 10% CD + 90% WW (C7), 30% CD + 70% WW (C6), 50% CD + 50% WW (C5), 70% CD + 30% WW (C4), 90% CD + 10% WW (C3), 100% CD (C2) and 100% WW (C1). The pH evolution and biogas (CH_4 and CO_2) production were followed for 25 days. Cultures tend to acidify with increase in cow dung proportion. Biogas production was significantly higher ($p < 0.05$) in C5 (880.0 mL), C6 (862.0 10 mL) and C7 (772.0 mL). Mixture C5 had a highest level of CO_2 production (40%). Also C7 and C6 retained in the experiment contained respectively organic matter, volatile fatty acids (VFA) and total alkalinity (TAC) as 41.06%, 47.02%, 1320 mg acetic acid/L, 3036 mg Acetic acid/L and 520 mg $CaCO_3$/L, 1310 mg $CaCO_3$/L. Mixture C6 was the best medium for microorganism proliferation stability with 3.5×10^5 UFC/ml of methanogens bacteria. It also possessed buffering capacity, which prevents acidification of medium during VFAs production.

Keywords

Bioconversion, Municipal Waste, Activated Sludge, Biomethane, Burkina Faso

1. Introduction

Demographic and economic growth, rapid urbanization and rising living standards lead to an increase in quantity and quality of solid waste production [1] [2] [3]. This phenomenon creates enormous risks to the environment and consequently to population health [4] [5]. The management of solid waste has become a major problem today, especially in developing countries (DCs), due to the lack of resources and difficulty to develop an approach adapted to their context. Despite recommendations and regulatory measures to reduce waste generation at source and ultimate waste, promoting eco-design, sorting, recycling and incineration, landfill remains a significant part of waste. In order to protect environment, the controlled methanization process in sealed enclosed spaces has been developed as opposed to landfill centers [6]. Indeed, the advantages of this technique are multiple. Methanization requires less space and considerably reduces volume and weight of waste to be buried. It permits significant reductions in greenhouse gas emissions (CO_2 and CH_4), and odors elimination. The digestate issued can be used to reinforce agriculture. The biogas (renewable fuel) generated can be used in several final applications. Therefore, methanization is more profitable than all other forms of waste treatment [7] [8] [9] [10] [11]. Biogas containing methane was recognized as a new renewable energy source according European Directive 2001/77/EC. Also, it added that 20% of energy consumed in EU in 2020 will coming from renewable sources. Thus, studies have been carried out in a context of anaerobic digestion process optimization. The different stages of anaerobic digestion process were dissected [12]-[17]. Many studies are interested in substrates diversification and their biomethanogenic potential [18] [19] [20] [21]. Co-digestion systems that treat a mixture of different king of waste, including animal, food and organic household waste have been developed [22] [23] [24]. Nations have embarked on energy crops to increase renewable energy production more rapidly [25] [26]. Amarante [27] reported several commercialized technologies worldwide for municipal organic residues treatment. The use of adequate inoculums is an important parameter for anaerobic digestion. It must be adapted to the substrate. Most digesters are inoculated with old sludges from other mesophilic or thermophiles anaerobic digesters [28] [29] [30]. In other studies a commercial solution containing some microorganisms was involved in anaerobic degradation [31]. Nikièma *et al.* [32] and Traoré *et al.* [33] showed the possibility of using wastewater and bovine dung from slaughterhouse as a source of inoculum for biogas production. Nature and proportion of inoculums used have potentially a significant impact on methane production [34] [35] [36] [37] [32]. The aim of this study was to develop activated sludge to inoculate municipal organic solid waste digesters.

2. Material and Methods

2.1. Inoculum Sampling

Wastewater (WW) and bovine dung (BD) were sampled at slaughterhouse in

Ouagadougou, Burkina Faso (12°25'5.87"N, 1°28'29.23"O). The cow dung was freshly removed, placed in a sterile stomacher plastic bag and transferred to the Laboratory. The wastewater was removed at depth (1 meter) and all samples were stored at 4°C.

2.2. Experimental Conditions

The experimental conditions for the production of activated sludge described was adapted from the process of Angelidaki [38]. The experimental conditions are presented in **Table 1**.

Experiments were carried out in 300 mL flask filled with 20 ml of mineral solution supplemented with 29 mL cellulose 5% (w/v). Combinations of different proportions of bovine dung and wastewater were used as inoculum: 100% WW, 100% CD, 90% CD + 10% WW, 70% CD + 30% WW, 50% CD + 50% WW, 30% CD + 70% WW and 10% CD + 90% WW. 0.4 ml of a buffer solution ($NaHCO_3$, 10%) and 0.6 ml of a reducing solution (Na_2S_9, H_2O, 2.4%) were added in a sterile manner. The tests were carried out in duplicate. The cow dung was washed in distilled water (2% w/v) prior to combinations.

The mineral solution contained 10 mL of phosphate buffer 10 mL (2 g/L K_2HPO_4 and 2 g/L NH_4Cl), 8 mL of Balch's trace mineral solution [39]; 1 mL of Widdel trace element solution [40] and 1 mL of vitamin solution.

2.3. Evaluation of Methanogenic Activities

The evaluation of activity of the different microbial groups was followed by the measurement of methane production [41]. Estimation of biogas production was carried out using liquid raising method. The gaseous products (CO_2 and CH_4) were analyzed using a gas chromatograph (Girdel series 30 catharometer equipped with a thermal conductivity detector [TCD] associated to SERVOTRACE potentiometer recorder type Sefram Paris of 1 mV). CH_4 and CO_2 were analyzed under following: Injector temperature 90°C, column temperature 60°C, detector temperature 100°C, filament current 150 mA, carrier gas pressure 1 bar, and attenuation 32, Paper flow rate 10 mm/min. 0.5 ml of the flask gas phase was taken

Table 1. Experimental design matrix.

Combination	MS (mL)	WW (mL)	BD (mL)	Cellulose 5% (mL)
100% WW (C1)	20	100	0	29
100% CD (C2)	20	0	100	29
90% CD + 10% WW (C3)	20	90	10	29
70% CD + 30% WW (C4)	20	70	30	29
50% CD + 50% WW (C5)	20	50	50	29
30% CD + 70% WW (C6)	20	30	70	29
10% CD + 90% WW (C7)	20	10	90	29

MS: Mineral solution, WW: Wastewater, CD: Cow dung.

and then injected into the chromatograph using a 1 ml waterproof syringe. The CH_4 and CO_2 content were determined using a standard curve based on CH_4 and CO_2.

2.4. Monitoring for the Selection of Microbial Complex

The microbial complex that gave more production of biogas was chosen as the best activated sludge. For this complex, the physicochemical parameters were determined. PH, salinity, Total dissolved solutes (TDS), electrical conductivity (EC) and resistivity were measured with a multi-parameter analyzer (9420 WTW). Dry matter (DM), total ash (Ct), organic matter (MO), total organic carbon (TOC) in the sample was obtained according to the protocol described by Sakaki [42]. Total alkalinity (TAC) and volatile fatty acids (VFAs) content were determined using the method described by Dilallo and Albertson [43]. An amount of 25 mL of the sample was titrated with H_2SO_4 (0.1 M) at pH 4.0 and the amount of titration was noted. The sample was then boiled lightly for 3 minutes, cooled, titrated with Na_2CO_3 (0.05 M), and the titration amount was noted from pH 4.0 to 7.0. Viable methanogenic bacteria were enumerated by the three-tube most probable number (MPN) technique (8-fold dilutions) using a medium adapted from Dianou [44] and a modified medium of Zhilina [45]. The composition of medium for serial dilution was respectively: NH_4Cl 2 g/L, K_2HPO4 0.33 g/L, KH_2PO_4 0.33 g/L, methanol 10 mM, sodium acetate 20 mM, sodium formate 20 mM, yeast extract 1 g/L, oligoelements solution 1 mL, resazurin solution (0.1%: v/v) 1 mL. Also 1 mL reducing solution was added to the medium and the headspace was filled with N_2 gas.

2.5. Statistical Analysis

The XLSAT software was used for statistical analysis of the data. The analysis of variance (ANOVA) was carried out to compare the mean values of biogas production obtained from different combinations using Fisher's tests at the probability threshold p = 5%. Principal Component Analysis (PCA) was performed to reduce geometric space and visualize data, using a linear combination of variables that maximizes variance. This method allowed to visualize the typology of the different combinations and to avoid the redundancy of the variables by considering the study in the reduced space of uncorrelated.

3. Results and Discussion

3.1. Evolution of pH

Figure 1 shows the evolution of pH according to the combinations of inoculums during the anaerobic digestion. The optimum pH during the anaerobic digestion is around neutrality and varied from, 6.5 to 8. pH value is less than 6.5 or greater than 8, a malfunction of the process [46] [47]. The analytical follow-up concerned the control of pH by adding NaOH (5N), in order to correct it. The pH of mixture C1 (100% WW) was increased to around 9. Parawina *et al.* [48] have

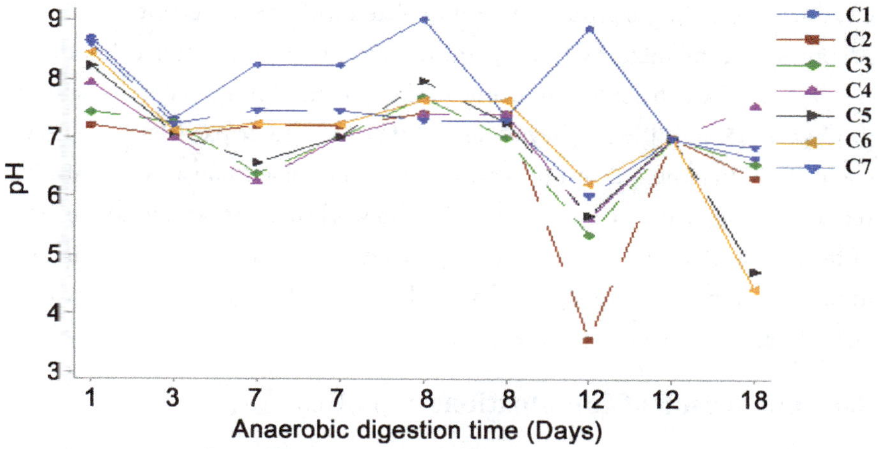

Figure 1. Evolution of the pH as a function of time of different inoculums.

found similar results with pH evolving from 6 to 9 until the 20th of anaerobic digestion. These results are in agreement with those previously reported by Cuetos *et al.* [49] and [50]. According to Bechir [51], wastewaters from the slaughterhouse of Ouagadougou are loaded with solid waste, organic material (residue of rumen) and concentrated by blood, flesh, fat, haires, excrement and urine of slaughtered animals. These waters have COD and BOD in order 2827, 1620 mg O_2/L, respectively [52]. This is confirmed by [53] [54]. Study of Cuetos *et al.* [49] was showed that the first 10 days, nitrogen compounds are degraded and ammonia nitrogen is released into the environment. Indeed, anaerobic bacteria mainly needed in organic form, nitrogen source (ammonia or urea nitrogen) and a source of phosphorus (orthophosphates). These nutrients are used for synthesis of molecule and energy during biological reactions. In the case of bovine dung, an increase of proportion causes acidification of environment. As also underlined some studies, the anaerobic digestion of bovine dung causes acidification of environment [55] [56]. Acidification is due to the presence of macromolecules in rapidly decomposing bovine dung [8] [57]. A weak variation in pH around neutrality (pH 7) was observed with the combinations C7 and C6.

3.2. Microbial Activities

Microbial activities were correlated by gas production during anaerobic digestion [41]. Studies showed that animal manure as co-substrate allows optimization of biomethane production, thus improving the anaerobic digestion system [58] [59]. These conditions permit a good development of microbial communities (bacteria and archaea) in anaerobic digestion process. Microbial community plays a role in process performance and stability [60].

3.3. Assessment of Biogas Production

Combinations C5, C6 and C7 were presented in **Table 2** and demonstrated no significant difference about biogas production. The difference in production was found significant ($p < 0.05$) between two combination groups as C5, C6, C7 and

C1, C4, C3, C2. Biogas production was depleted with the increasing of cow dung proportion in the mixture. In fact, cow dung contains macromolecules, the disintegration of which could produce acids components implicated in instability of process [8] [55] [57] [61]. They argument that some toxic particles or molecules could limit anaerobic digestion and consequently block biogas synthesis. This report was supported by Mekonnen [57] who worked on anaerobic co-treatment of tannery wastewater and bovine dung. It revealed that tanneries rich in chromium and sulphide, were highly toxic to the anaerobic digestion process causing fall of biogas production [57] [62].

3.4. Comparison of Combinations Activities

The influence of matrices characteristics on the selection of combination was based on principal components analysis (PCA). This analysis was used for description and visualization of combinations with three parameters studied: Biogas, Methane (CH_4) and Carbon dioxide (CO_2). The main components are shown in **Figure 2**. The matrix of linear correlations is shown in **Table 3**. The Kaiser criterion indicated that two (02) principal components (PC) should be considered. Indeed, the three (02) first main components have an Eigen value greater than 1 (**Figure 2**). However, PC3 accounts for only 0.75% of data variability (**Figure 2**). Therefore, only the first two (02) components describing 99%

Table 2. Biogas production after 25 days of anaerobic digestion (mean of 3 replicates).

Inoculum	Volume of biogas (mL)
C5	880.0[a]
C6	862.0[a,b]
C7	772.0[a,b]
C1	668[b,c]
C4	533.0[c,d]
C3	414.0[d]
C2	343.0[d]

In a column, values that have a different letter are not significantly different according to the Fisher test (LSD) at 5% threshold.

Table 3. Coordinates of variables and their contribution to identification.

Variables	Principal components		
	F1	F2	F3
Biogas (mL)	0.570	0.587	−0.575
CH_4 (%)	0.755	−0.097	0.649
CO_2 (%)	−0.325	0.803	0.499
Eigen values	1.719	1.259	0.023
Variance (%)	57.292	41.952	0.756
Cumulative (%)	57.292	99.244	100.000

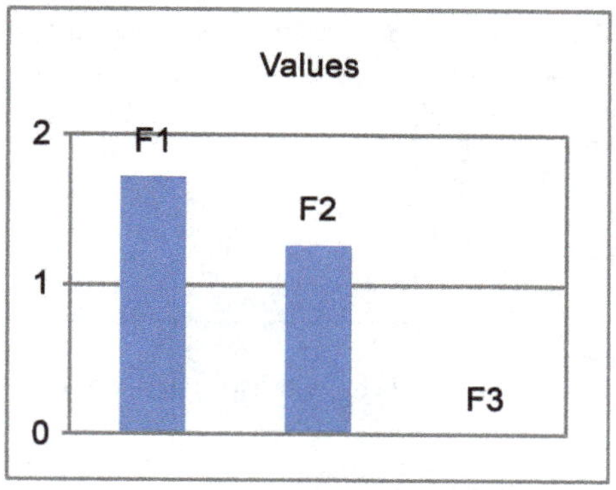

Figure 2. Results of principal component analysis.

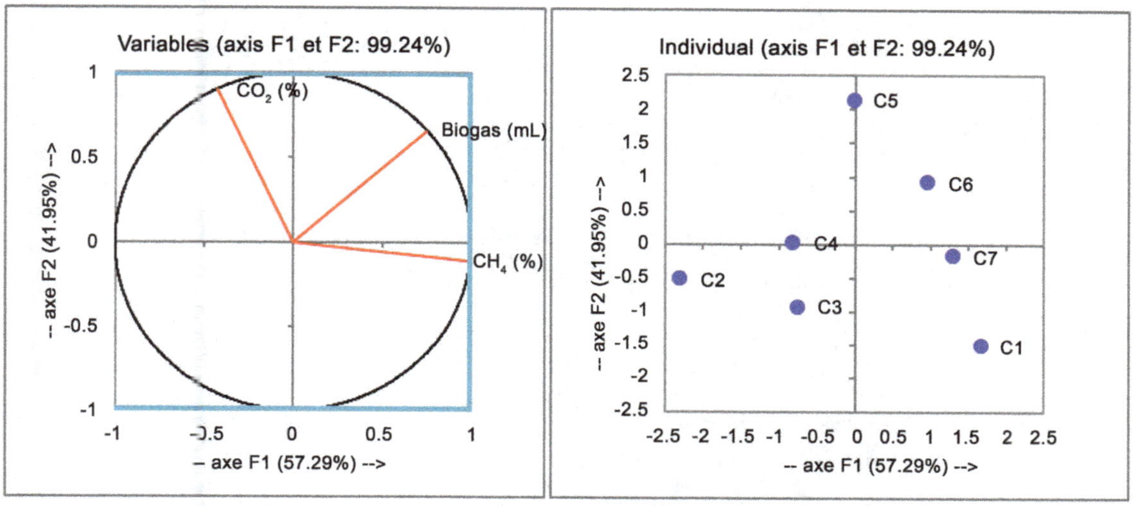

(a) (b)

Figure 3. Principal component analysis (a) plot of variables biogas production, CO_2 and CH_4 proportions and (a) distribution of combinations on 1 × 2 axis of principal components.

of the data variability were retained for data description.

The typologies of variables (Biogas, CH_4 and CO_2) and samples on factorial planes constituted by axes 1 and 2 were presented in **Figure 3(a)** and **Figure 3(b)**, respectively. In these two figures, only variables close to correlation circle should be taken into account. In **Figure 3(a)** there are clearly three groups of variables close to the circle (Biogas, CH_4 and CO_2, so that the projections on the F1 and F2 axes are 99.24%). The first group consisting of the Biogas variable was positively correlated with the two axes F1 and F2. The second group (CH_4) was positively correlated with axis 1 and negatively with axis 2. The third group (CO_2) was negatively correlated with axis 1 and positively with axis 2. It is interesting to note that there is no particular affinity or opposition with the production of Biogas and the proportions of CH_4 and CO_2 (**Table 4**). Since the biogas consists of CH_4 and CO_2, the correlation exists between the productions of different gas. The representation of the combinations on the two factorial planes described by the axes F1 × F2 (**Figure 3(a)**) makes it possible to compare them according to the production of biogas and the proportions in CH_4 and CO_2. Combinations C5, C7 and C6 have a significant production of biogas. Combina-

Table 4. Pearson correlation matrix of variables distributions.

Principal components	Biogas (mL)	CH_4 (%)	CO_2 (%)
Biogas (mL)	1	0.659	0.269
CH_4 (%)		1	−0.513
CO_2 (%)			1

Table 5. Physicochemical characteristics of the best microbial complex.

Parameter	Means values	
	C7	C6
pH	7.21	7.68
TS (g/L)	10.52	17.48
TSV (%)	41.06	47.02
Ash (%)	58.93	52.97
COT (%)	23.81	27.54
Salinity (g/L)	3.5	7
TDS (g/L)	6.33	12.27
EC (mS/cm)	6.32	12.27
R (Ω*cm)	158.1	81.6
VFAs (mg acetic acid/L)	1320	3036
TAC (mg $CaCO_3$ /L)	520	1310
Methanogenic bacteria (CFU/mL)	3.5×10^5	3.5×10^5

CD = cow dung; WW = wastewater; C7 = 10CD + 90WW; C6 = 30CD + 70WW; TDS = Total dissolved solutes; EC = Electrical conductivity; R = Resistivity.

tions C1, C7 and C6 have high proportions of CH_4. And the combination C5 has a high proportion of CO_2.

3.5. Characteristics of the Sludge Produced

The physicochemical characteristics of the C7 and C6 sludge were presented in **Table 5**. Organic matter was respectively 41.06%, 47.02% for C7 and C6. VFAs and TAC concentration were 1320 mg acetic acid/L, 520 mg $CaCO_3$/L for C7 and 3036 mg acetic acid/L, 1310 mg $CaCO_3$ /L for C6. These characteristics are comparable to those of [50] [63] [64] [65] [66]. Ali Shah [67] showed that physicochemical parameters influence the densities of archaea and bacteria in the medium. In died, according to Aguilar [68] significant inhibitory effect was observed in the process at 10,000 mg/L of VFAs. Vedrenne [69] was found 2000 - 3000 mg/L with the same substrate. The alkalinity of the digesters should not exceed 1000 mg of $CaCO_3$/l of alkalinity according to Hawkes *et al.* [70]. The VFAs values in the case of C6 allow adaptation to microorganisms with high values of acid production in the digesters. Methanogenic bacteria were estimated to 3.5×10^5 CFU/mL for C7 and C6. Ueki *et al.* [71] obtained respectively 7.17 ×

10^4/mL or 1.6×10^6/mL with H_2 or acetate as the substrate, in anaerobic digestion of municipal sewage sludge.

4. Conclusion

The evolution of the pH showed that the formulation with the low proportion in cow dung had the best production profile. C7 and C6 gave best production of CH_4 and CO_2 then also exhibited the growth of methanogenic bacteria. Their physico-chemical characteristics showed an interesting media rich in organic matter, VFAs and TAC which could maintain microbial flora stability. The combination C6 was presented good conditions of growth and maintenance associated to anaerobic digestion of bacterial consortia. It buffering properties can use to prevent medium acidification during VFAs production. To conclude C6 could be considered as best inoculum for organic waste digesters.

Acknowledgements

We thank the Cooperation and Cultural Action Service (SCAC) of the French Embassy in Burkina Faso and UEMOA through the CRSBAN/University Ouaga I Prof. Joseph KI-ZERBO for their financial support.

References

[1] Alouemine, S.O. (2006) Méthodologie de caractérisation des déchets ménagers à Nouakchott (Mauritanie): Contribution à la gestion des déchets et outils d'aide à la décision. Thèse, Université de Limoges, 192 p.

[2] Guerrero, A.L., Maas, G. and Hogland, W. (2013) Waste Management. *Waste Management,* **33**, 220-232. https://doi.org/10.1016/j.wasman.2012.09.008

[3] Minghua, Z., Xiumin, F., Rovetta, A., Qichang, H., Vicentini, F., Bingkai, L., Giusti, A. and Yi, L. (2009) Municipal Solid Waste Management in Pudong New Area, China. *Waste Management,* **29**, 1227-1233. https://doi.org/10.1016/j.wasman.2008.07.016

[4] Sharholy, M., Ahmad, Vaishya, K.R.C. and Gupta, R.D. (2007) Municipal Solid Waste Characteristics and Management in Allahabad, India. *Waste Management,* **27**, 490-496. https://doi.org/10.1016/j.wasman.2006.03.001

[5] Seng, B., Kaneko, H., Hirayama, K. and Katayama-hirayama, K. (2011) Municipal solid Waste Management in Phnom Penh, Capital City of Cambodia. *Waste Management & Resource,* **29**, 5. https://doi.org/10.1177/0734242X10380994

[6] Directive Landfill (1999) Council Directive 1999/31/EC of 26 the April 1999 on the landfill of waste. *Official Journal of the European Communities,* L 18/1-19.

[7] Romano, P.V., Krogmann, U., Westendorf, M.L. and Strom, P.F. (2006) Small-Scale Composting of Horse Manure Mixed with Wood Shavings Small-Scale Composting of Horse Manure Mixed with Wood Shavings. *Compost Science & Utilization,* **14**, 132-141. https://doi.org/10.1080/1065657X.2006.10702274

[8] Sakar, S., Yetilmezsoy, K. and Kocak, E. (2009) Anaerobic Digestion Technology in

Poultry and Livestock Waste Treatment—A Literature Review. *Waste Management and Research*, **27**, 3-18. https://doi.org/10.1177/0734242X07079060

[9] M'Sadak, Y., Ben M'Barek, A. and Baraket, S. (2011) Suivis physico-chimique et énergétique de la biométhanisation expérimentale appliquée à la biomasse bovine. *Revue "Nature et Technology"*, n° 07, 81-86.

[10] Nelson, M.C., Morrison, M., Schanbacher, F. and Yu, Z. (2012) Shifts in Microbial Community Structure of Granular and Liquid Biomass in Response to Changes to Infeed and Digester Design in Anaerobic Digesters Receiving Food-Processing Wastes. *Bioresource Technology*, **107**, 135-1343. https://doi.org/10.1016/j.biortech.2011.12.070

[11] Appels, L., Houtmeyers, S., Ruyters, S., Busschaert, P., Lievens, B., Impe, J.V. and Dewi, R. (2013) Influence of Sludge Pre-Treatment on the Microbial Community Structure in Anaerobic Digesters. *13th World Congr Anaerob Dig.*

[12] Zehnder, A.J.B. and Brock, T.D. (1979) Methane Formation and Methane Oxidation by Methanogenic Bacteria. *Journal of Bacteriology*, **137**, 420-432.

[13] Stams, A.J.M. (1994) Metabolic Interactions between Anaerobic Bacteria in Methanogenic Environments. *Antonie Van Leeuwenhoek*, **66**, 271-294. https://doi.org/10.1007/BF00871644

[14] Lyberatos, G. and Skiadas, I.V. (1999) Modelling of Anaerobic Digestion—A Review. *GlobalNEST International Journal*, **1**, 63-76.

[15] Ostrem, K. (2004) Greening Waste: Anaerobic Digestion for Treating the Organic Fraction of Municipal Solid Waste. Columbia University (The Earth Engineering Center and the Henry Krumb School of Mines), 1-59.

[16] Vigneron, V. (2005) Voies de réduction des oxydes d'azote lors de leur injection dans un massif de déchets ménagers et assimilés. Thèse, Université Paris-XII, Val-de-Marne.

[17] Amani, T., Nosrati, M. and Sreekrishnan, T.R. (2010) Anaerobic Digestion from the Viewpoint of Microbiological, Chemical, and Operational Aspects—A Review. *Environmental Reviews*, **18**, 255-278. https://doi.org/10.1139/A10-011

[18] Girault, R., Peu, P., Béline, F., Lendormi, T. and Guillaume, S. (2013) Caractéristiques des substrats et interactions dans les filières de co-digestion: Cas particulier des co-substrats d'origine agro-industrielle. *Sciences Eaux & Territoires*, **3**, 44-53.

[19] Barrantes, L.M., Hosseini, K.E. and Eskicioglu, C. (2014) Anaerobic Co-Digestion of Wine/Fruit-Juice Production Waste with Landfill Leachate Diluted Municipal Sludge Cake under Semi-Continuous Flow Operation. *Waste Management*, **34**, 1860-1870. https://doi.org/10.1016/j.wasman.2014.06.027

[20] Hosseini, K.E., Barrantes, L.M., Eskicioglu, C. and Dutil, C. (2014) Mesophilic Batch Anaerobic Co-Digestion of Fruit-Juice Industrial Waste and Municipal Waste Sludge: Process and Cost-Benefit Analysis. *Bioresource Technology*, **152**, 66-73. https://doi.org/10.1016/j.biortech.2013.10.072

[21] Zhang, Q., Hu, J. and Lee, D. (2016) Biogas from Anaerobic Digestion Processes: Research Updates. *Renew Energy*, **98**, 108-119. https://doi.org/10.1016/j.renene.2016.02.029

[22] Hartmann, H. and Ahring, B.K. (2005) Anaerobic Digestion of the Organic Fraction of Municipal Solid Waste: Influence of Co-Digestion with Manure. *Water Resource*, **39**, 1543-1552.

[23] Macias-Corral, M., Samani, Z., Hanson, A., Smith, G., Funk, P., Yu, H. and Longworth, J. (2008) Anaerobic Digestion of Municipal Solid Waste and Agricultural Waste and the Effect of Co-Digestion with Dairy Cow Manure. *Bioresource Tech-*

nology, 99, 8288-8293.
https://doi.org/10.1016/j.biortech.2008.03.057

[24] Nielsen, H.B. and Angelidaki, I. (2008) Congestion of Manure and Industrial Organic Waste at Centralized Biogas Plants: Process Imbalances and Limitations. *Water Science Technology*, **58**, 1521-1528. https://doi.org/10.2166/wst.2008.507

[25] Rsoud, M.B. (1999) Développement durable et analyse énergétique des exploitations agricoles. *Économie Rural*, **252**, 16-27.
https://doi.org/10.3406/ecoru.1999.5096

[26] Rsoud, B. and Chopinet B. (1999) Efficacité énergétique et diversité des systèmes de production agricole—Application à des exploitations bourguignonnes. Ingénieries-EAT, IRSTEA Edition 1999, 17-25.

[27] Amarante, J.A.L. (2010) Biométhanisation des déchets putrescibles municipaux—Technologies disponibles et enjeux pour le Québec. 99.

[28] Shelton, D.R. and Tiedjei, J.M. (1984) Isolation and Partial Characterization of Bacteria in an Anaerobic Consortium That Mineralizes 3-Chlorobenzoic Acid. *Applied Environmental Microbiology*, **48**, 840-848.

[29] Chynoweth, D.P., Owens, J.M. and Legrand, R. (2001) Renewable Methane from Anaerobic Digestion of Biomass. *Renew Energy*, **22**, 1-8.
https://doi.org/10.1016/S0960-1481(00)00019-7

[30] Hansen, T.L., Ejbye, J., Angelidaki, I., Marca, E., Jansen, C., Mosbæk, H. and Christensen, T.H. (2004) Method for Determination of Methane Potentials of Solid Organic Waste. *Waste Management*, **24**, 393-400.
https://doi.org/10.1016/j.wasman.2003.09.009

[31] Lee, H. and Shoda, M. (2008) Stimulation of Anaerobic Digestion of Thickened Sewage Sludge by Iron-Rich Sludge Produced by the Fenton Method. *Journal of Bioscience and Bioengineering*, **106**, 107-110. https://doi.org/10.1263/jbb.106.107

[32] Nikiema, M., Sawadogo, J.B., Somda, M.K., Traore, D., Dianou, D. and Traore, A. (2015) Optimisation de la production de biométhane à partir des déchets organiques municipaux [Optimization of Biomethane Production from Municipal Solid Organic Wastes]. *International Journal of Biological and Chemical Sciences*, **9**, 2743-2756. https://doi.org/10.4314/ijbcs.v9i5.43

[33] Traoré, D., Dianou, D. and Traoré, A.S. (2017) Valorization of Mango Waste by Biomethane Production in Burkina Faso. *Current Research in Microbiology and Biotechnology*, **5**, 1212-1218.

[34] Owen, W.F., Stuckev, D.C., Healv, J.B., Young, L.Y. and Mccagrv, P.L. (1979) Bioassay for Monitoring of Biochemical Methane Potential and Anaerobic Toxicity. *Water Resource*, **13**, 485-492.

[35] Lopes, S.W., Leite, D.V. and Prasad, S. (2004) Influence of Inoculum on Performance of Anaerobic Reactors for Treating Municipal Solid Waste. *Bioresource Technology*, **94**, 261-266. https://doi.org/10.1016/j.biortech.2004.01.006

[36] Bouez, M. (2008) Dégradation anaérobie de déchets solides: Caractérisation, facteurs d'influence et modélisations. Thèse, 259 p.

[37] Park, H., Rosenthal, A., Jezek, R., Ramalingam, K., Fillos. J. and Chandran, K. (2010) Impact of Inocula and Growth Mode on the Molecular Microbial Ecology of Anaerobic Ammonia Oxidation (Anammox) Bioreactor Communities. *Water Resource*, **44**, 5005-5013. https://doi.org/10.1016/j.watres.2010.07.022

[38] Angelidaki, I., Alves, M., Bolzonella, D., Borzacconi, L., Campos, J.L., Guwy, A.J., Kalyuzhnyi, S., Jenicek, P. and van Lier, J.B. (2009) Defining the Biomethane Poten-

tial (BMP) of Solid Organic Wastes and Energy Crops: A Proposed Protocol for Batch Assays. *Water Science & Technology*, **59**, 927-934. https://doi.org/10.2166/wst.2009.040

[39] Balch, W.E., Fox, G.E., Magrum, L.J., Woese, C.R. and Wolfe, R.S. (1979) Methanogens: Reevaluation of a Unique Biological Group. *Microbiology Reviews*, **43**, 260-296. http://www.ncbi.nlm.nih.gov/pmc/articles/PMC281474/

[40] Widdel, N. and Pfennig, F. (1981) Studies on Dissimilatory Sulfate-Reducing Bacteria That Decompose Fatty Acids. *Archives of Microbiology*, **129**, 395-400. https://doi.org/10.1007/BF00406470

[41] Nopharatana, A., Clarke, W.P., Pullammanapallil, P.C., Silvey, P. and Chynoweth, D.P. (1998) Evaluation of Methanogenic Activities during Anaerobic Digestion of Municipale Solid Waste. *Bioresource Technology*, **64**, 169-174. https://doi.org/10.1016/S0960-8524(97)00190-9

[42] Sakaki, S. (2014) Etude de la production des acides gras volatils à partir de la fermentation acidogène des boues d'épuration des effluents issus d'un e usine de pâtes et papiers de la fermentation acidogène des boues d'épuration des effluents issus d'une usine de pates et papiers. Thèse, Université du Québec, Québec, 84.

[43] DiLallo, R. and Albertson, O.E. (2014) Volatile Acids by Direct Titration. *Water Environment*, **33**, 356-365.

[44] Dianou, D. (1993) Etude de l'influence des interactions entre les bactéries réductrices du cycle du soufre et les bactéries méthanogéniques sur la production du riz des bas-fonds. Thèse, Université de Ouagadougou, Ouagadougou, 155.

[45] Zhilina, T.N. (1972) Death of Methanosarcina in the Air. *Microbiology*, **40**, 980-981.

[46] Lee, D.H., Behera, S.K., Kim, J.W. and Park, H. (2009) Methane Production Potential of Leachate Generated from Korean Food Waste Recycling Facilities: A Lab-Scale Study. *Waste Management*, **29**, 876-882. https://doi.org/10.1016/j.wasman.2008.06.033

[47] Dobre, P., Nicolae, F. and Matei, F. (2014) Main Factors Affecting Biogas Production—An Overview. *Romanian Biotechnological Letters*, **19**, 9283-9296.

[48] Parawira, W., Murto, M., Zvauya, R. and Mattiasson, B. (2006) Comparative Performance of a UASB Reactor and an Anaerobic Packed-Bed Reactor When Treating Potato Waste Leachate. *Renew Energy*, **31**, 893-903. https://doi.org/10.1016/j.renene.2005.05.013

[49] Cuetos, M.J., Gomez, X., Otero, M. and Moran A. (2008) Anaerobic Digestion of Solid Slaughterhouse Waste (SHW) at Laboratory Scale: Influence of Co-Digestion with the Organic Fraction of Municipal Solid Waste (OFMSW). *Biochemical Engineering Journal*, **40**, 99-106. https://doi.org/10.1016/j.bej.2007.11.019

[50] Bouallagui, H., Rachdi, B., Gannoun, H. and Hamdi, M. (2009) Mesophilic and Thermophilic Anaerobic Co-Digestion of Abattoir Wastewater and Fruit and Vegetable Waste in Anaerobic Sequencing Batch Reactors. *Biodegradation*, **20**, 401-409.

[51] Bechir, M. (1989) Etude de la réhabilitation de la station de traitement des eaux usées de l'abattoir de Ouagadougou. 101.

[52] Zongo, I., Diallo-Kone, M., Palm, K., Tiemtore, A., Sanogo, O., Guiguemde, E., Lapicque, F. and Leclerc, J.P. (2012) Review of Wastewater from the City of Ouagadougou: Self-Purification Capacity for the Production of Biogas. *Study & Research: Chemistry & Chemical Engineering, Biotechnology, Food Industry*, **13**, 153-167.

[53] Asselin, M. (2007) Utilisation de l'électrocoagulation dans le traitement des eaux

usées industrielles chargées en matières organique. Université du Québec, Québec.

[54] Khennoussi, A., Chaouch, M. and Chahlaoui, A. (2013) Traitement des effluents d'abattoir de viande rouge par électrocoagulation flottation avec des électrodes en fer. *Revue des Sciences de l'Eau*, **26**, 81-171. http://id.erudit.org/iderudit/1016064ar

[55] Zhu, G. and Jha, A.K. (2013) Psychrophilic Dry Anaerobic Digestion of Cow Dung for Methane Production: Effect of Inoculum. *Science Asia*, **39**, 500-510. https://doi.org/10.2306/scienceasia1513-1874.2013.39.500

[56] Fatima, K.C. and Ale, B.B. (2015) Production of Biogas from Slaughterhouse Waste in Lalitpur Sub-Metropolitan City. *Proceedings of IOE Graduate Conference*, 143-149.

[57] Mekonnen, A., Leta, S. and Njau, N.K. (2017) Kinetic Analysis of Anaerobic Sequencing Batch Reactor for the Treatment of Tannery Wastewater. *African Journal of Environment Sciences and Technology*, **11**, 339-348. http://academicjournals.org/journal/AJEST/article-abstract/022BE1B64383 https://doi.org/10.5897/AJEST2017.2305

[58] Afilal, M.E., Belkhadi, N., Daoudi, H. and Elasri, O. (2013) Methanic Fermentation of Different Organic Substrates. *Journal of Materials and Environmental Science*, **4**, 11-16.

[59] Chomini, M.S., Ogbonna, C.I.C., Falemara, B.C. and Micah, P. (2015) Effect of Co-Digestion of Cow Dung and Poultry Manure on Biogas Yield, Proximate and Amino Acid Contents of Their Effluents. *Journal of Agriculture and Veterinary Science*, **8**, 2319-2372.

[60] Wang, S., Hou, X. and Su, H. (2017) Exploration of the Relationship between Biogas Production and Microbial Community under High Salinity Conditions. *Scientific Reports*, **7**, Article No. 1149. http://www.nature.com/articles/s41598-017-01298-y https://doi.org/10.1038/s41598-017-01298-y

[61] Abubakar, B.S.U.I and Ismail, N. (2017) Anaerobic Digestion of Cow Dung for Biogas Production. *Journal of Engineering and Applied Sciences*, **7**, 3-7.

[62] Mondal, N.C., Saxena, V.K. and Singh, V.S. (2005) Assessment of Groundwater Pollution Due to Tannery Industries in and around Dindigul, Tamilnadu, India. *Environmental Geology*, **48**, 149-157. https://doi.org/10.1007/s00254-005-1244-z

[63] Banks, C.J. and Wang, Z. (1999) Development of a Two Phase Anaerobic Digester for the Treatment of Mixed Abattoir Wastes. *Water Science & Technology*, **40**, 69-76.

[64] Yi, J., Dong, B., Jin, J. and Dai, X. (2014) Effect of Increasing Total Solids Contents on Anaerobic Digestion of Food Waste Under Mesophilic Conditions: Performance and Microbial Characteristics Analysis. *PLoS ONE*, **9**, e102548. https://doi.org/10.1371/journal.pone.0102548

[65] Dhamodharan, K., Kumar, V. and Kalamdhad, A.S. (2015) Effect of Different Livestock Dungs as Inoculum on Food Waste Anaerobic Digestion and Its Kinetics. *Bioresource Technology*, **180**, 237-241. https://doi.org/10.1016/j.biortech.2014.12.066

[66] Rajagopal, R. and Massé, D.I. (2016) Start-Up of Dry Anaerobic Digestion System for Processing Solid Poultry Litter Using Adapted Liquid Inoculum. *Process Safety and Environmental Protection*, **102**, 495-502. https://doi.org/10.1016/j.psep.2016.05.003

[67] Ali Shah, F., Mahmood, Q., Maroof Shah, M., Pervez, A. and Ahmad Asad, S. (2014) Microbial Ecology of Anaerobic Digesters: The Key Players of Anaerobiosis. *The Scientific World Journal*, **2014**, Article ID: 183752.

[68] Aguilar, A., Casas, C. and Lema, J.M. (1995) Degradation of Volatile Fatty Acids by Differently Enriched Methanogenic Cultures: Kinetics and Inhibition. *Water Research*, **29**, 505-509. https://doi.org/10.1016/0043-1354(94)00179-B

[69] Vedrenne, F., Beline, F. and Bernet, N. (2007) Etude des processus de dégradation anaérobie et de production de méthane au cours du stockage des lisiers. Vol. Doctorat, Vie, Agronomie et Santé.

[70] Hawkes, F.R., Guwy, A.J., Rozzi, A.G. and Hawkes, D.L. (1993) A New Instrument for On-Line Measurement of Bicarbonate Alkalinity. *Water Research*, **27**, 167-170. https://doi.org/10.1016/0043-1354(93)90208-Y

[71] Ueki, K., Ueki, K., Takahashi, K. and Iwatsu, M. (1992) The Role of Sulfate Reducing in Methanogenic Digestion of Municipal Sewage Sludge. *Journal of General and Applied Microbiology*, **38**, 195-207. https://doi.org/10.2323/jgam.38.195

9

Modeling Plume-Rise of Air Emissions from Animal Housing Systems: Inverse AERMOD

Manqing Ying[1], Lingjuan Wang-Li[1*], Larry F. Stikeleather[1], Jack Edwards[2]

[1]Department of Biological and Agricultural Engineering, North Carolina State University, Raleigh, USA
[2]Department of Mechanical and Aerospace Engineering, North Carolina State University, Raleigh, USA
Email: *lwang5@ncsu.edu

Abstract

As fate and transport of air emissions from animal housing systems is of increasing concern, dispersion models have become commonly used tools to estimate the downwind concentrations of pollutants at certain locations surrounding the animal production farms. In application of Gaussian dispersion model for downwind concentration predictions of animal housing emissions, unknown plume rise (Δh) and plume shape of the horizontally emitted plumes from animal housing systems have been vital weak points challenging the accuracy of the model predictions. This paper reports an inverse AERMOD modeling study to derive the plum rises of PM_{10} emissions from mechanically ventilated egg production houses based upon field measurements of PM_{10} emission rate, downwind concentrations, and meteorological conditions. In total, 87 hourly plume rises were found for 20 days (five days per season for four seasons, from fall 2008 to summer 2009). The mean plume rises for fall 2008, winter 2008, spring 2009 and summer 2009 were 16.2 m (SE = 11.2 m), 7.9 m (SE = 9.5 m), 16.5 m (SE = 12.4 m), and 14.3 m (SE = 10.0 m), respectively. The relationships between plume rises and various factors were tested. While the diurnal patterns of the plume rises were not consistent among different selective days, they generally followed the diurnal patterns of house ventilation rates. Plume rise for weekends were significantly higher than those for weekdays in fall. Multiple linear regression showed a significant positive relationship ($p = 0.0134$) between wind speed and the plume rises. Ambient relative humidity and total volume flow were also found to be slightly ($p = 0.171$ and 0.217, respectively) related to the plume rises.

Keywords

Animal Feeding Operation, Air Emission, Plume-Rise, AERMOD

1. Introduction

Animal feeding operations (AFOs) emit significant amount aerial pollutants, contributing to the elevated air pollutant concentrations in the vicinity areas [1]-[9]. Although air emissions from AFOs are of great environmental concern in the United States (U.S.), application of the current National Ambient Air Quality Standards (NAAQS) is challenged due to lack of knowledge about air dispersions of AFO air emissions. Fate and transport of air emissions from AFO facilities have become a growing political and environmental concern [10] [11] [12].

In the studies of AFOs air emissions, atmospheric dispersion models are often used to predict downwind concentrations of the emitted pollutants [13]-[19], or, to derive emission rates through inverse modeling [20]-[26]. Among various types of dispersion models (e.g., Box models, Gaussian models, Lagrangian and Eulerian models, and CFD model) [27], Gaussian plume models are the most widely used for numerically describing dispersion of air emissions from different types of sources (e.g., point, line, area, volume; or, industrial stacks versus agricultural operations; etc.) [28] [29] [30] [31]. In application of the Gaussian dispersion models, knowledge about plume-rise is required for simulating the dispersion of the plume and calculating the concentrations at downwind locations [28] [32] [33] [34]. By definition, plume rise is the rise of the dispersing plume centerline above its original emission source height (*i.e.*, the physical stack height). In the literature, dispersion plume rises from various emission sources have been experimentally studied [35]-[42] and several plume rise models were developed from observations of emission plumes from industrial stacks (**Figure 1(a)** & **Figure 1(b)**) with both vertical momentum and thermal buoyancy [32] [43] [44] [45] [46] [47]. In modern animal production houses, mechanical ventilation systems are used to create optimal growing conditions for animals. Aerial pollutants from those houses are usually released through ventilation fans at ground level and horizontal positions or positions slightly tipping downwards (**Figure 1(c)**). There is no actual emission "stack", nor is there upward vertical velocity. As a result, the current plume rise models do not apply for AFO air emission dispersion modeling. Unknown plume-rise and plume shape (*i.e.*, horizontal and vertical plume spread parameters) are vital weak points challenging the accuracy of the Gaussian model predictions.

Figure 1. Visual comparison of various emission plumes. (a) Power plant stacks, Source: http://www.epa.gov/airmarkets/participants/monitoring/index.html; (b) Industrial stack; (c) Emission plume from two tunnel ventilated poultry houses (image by Wang-Li).

As a preferred regulatory air dispersion model of the U.S. Environmental Protection Agency (EPA), AERMOD was developed by the American Meteorological Society (AMS) and EPA Regulatory Model Improvement Committee (AERMIC) in early 1990's. This AERMIC Model (*i.e.*, AERMOD) was a replacement of the previous regulatory model ISCST3. Like ISCST3, AERMOD is a Gaussian-based model that incorporates advanced understanding of the boundary theory and turbulence with inclusion of terrain interactions [48] [49] [50]. In AERMOD, at stable conditions, the plume dispersion is treated as Gaussian distribution in both vertical and horizontal directions; the plume rise is determined by Briggs equations [48] [49] that takes temperature and wind changes above stack top into consideration. At unstable conditions (convective conditions), the plume dispersion is treated as Gaussian distribution in horizontal direction and as bi-Gaussian probability distribution in vertical direction; the plume rise is determined using convective plume rise formulations that account for convective updrafts and downdrafts [48] [49]. Although AERMOD is far more advanced than other Gaussian-based models, it still shares the deficiency in assessing the plum rise for emission sources with ground level and horizontal emission characteristics as it is of animal housing emission. Since AERMOD has been adopted by the State Air Pollution Regulatory Agencies (SAPRAs) to predict downwind concentrations or to derive emission rates through inverse modeling, and/or to compile emission inventories for agricultural emission sources, it is essential to develop understand the dynamics of the plume rises of agricultural air emissions as impacted by the production practices and atmospheric conditions. Therefore, the objectives of this research were to apply inverse-AERMOD to derive the plume rises of animal house air emissions using field measurements of particulate matter (PM) emissions and downwind concentrations; and to identify the impacts of various factors on plume rises from the animal houses based upon the inverse-AERMOD results and the field observations.

2. Materials and Methods

In this study, plume rises for animal housing systems were predicted using inverse AERMOD approach. The PM_{10} dispersion from a commercial egg production farm was modeled based upon measured emission rates. The plume rises for the source types (*i.e.*, ventilation types) were adjusted and the results of model predicted hourly downwind PM_{10} concentrations were compared with field observations at four ambient stations in the vicinity. When the model prediction agreed with the field measurement within ±20% relative difference, the plume rise for that given hour was considered valid. In total, 480 hourly PM_{10} data for four seasons with five days per season for four seasons were selected for this modeling study.

2.1. The Egg Production Farm and the PM_{10} Monitoring Station

The PM data collection was conducted on a commercial egg production farm in

North Carolina. As shown in **Figure 2**, this farm consisted of nine layer houses and an egg packing building. Based on the house type and ventilation type, the production houses were grouped into (1 tunnel ventilated high-rise layer houses (houses 1 - 4); (2 cross ventilated high-rise layer houses (houses 5 - 6); and (3 naturally ventilated shallow pit layer houses (houses 93, 102 - 103). In the high-rise houses, laying hens were caged on the second floor in 6 rows (houses 1 - 4) or 8 rows (houses 5 - 6) and manure was stored in the first floor and was completely cleaned out once a year. In the naturally ventilated houses, hens were caged in 4 tires and manure was flushing out to lagoons twice a day. The dimensions of the houses are listed in **Table 1**.

The tunnel ventilated houses used 17 belt-driven ventilation fans (upper 8 and lower 9, 1.22 m in diameter) on end wall of each house (**Figure 3**). The cross-ventilated houses had 27 belt-driven ventilation fans (1.22 m in diameter) arrayed on each side wall of each house (**Figure 3**). The naturally ventilated had curtain openings on the side walls, and twenty winter vents evenly distributed on the building ridge.

The PM concentrations were monitored at 5 locations, of which one was the in-house monitoring station (ST1 in **Figure 2**) and 4 were ambient monitoring

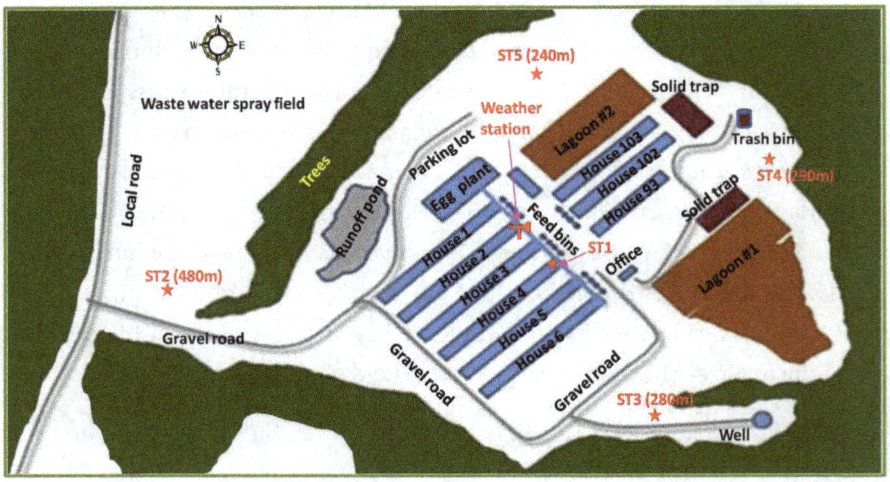

ST1 = station 1 (inside the house 4); ST2 = station 2; ST3 = station 3; ST4 = station 4; ST5 = station 5

Figure 2. The egg production farm layout and the PM_{10} monitoring stations [51].

Table 1. Building information of the production houses.

House	Length (m)	Width (m)	Height (m)	Ventilation type	Hen population
1 - 4	176.8	17.8	5.5 (eave) 8.4 (ridge)	Tunnel	~96,000 per house
5 - 6	176.8	21.0	5.5 (eave) 8.4 (ridge)	Cross	~130,000 per house
102 - 103	152.4	17.8	3.0 (eave) 5.5 (ridge)	Nature	~70,000 per house
93	107.0	17.8	2.5 (eave) 5.0 (ridge)	Nature	~30,000 per house

Figure 3. Illustration of the fans in tunnel-ventilated houses (top) & a cross-ventilated house (bottom).

stations (ST2-5 in **Figure 2**). The ST1 was designed to measure in-house PM concentration for determination of PM emission rate from the tunnel ventilated houses, whereas ST2-5 were designed to simultaneously measure downwind and upwind PM concentrations. Also, there is another PM monitor at the ventilation air inlet of house 4, measuring the background PM concentrations of the house inlet air. At each PM monitoring station, one tapered element oscillating microbalance (TEOM) equipped with a federal reference method PM_{10} sampling head was used to take continuous PM_{10} measurements [5] [51]. A 10 m weather tower with sensors for solar radiation, wind speed and direction, ambient temperature and relative humidity (RH) were used to take onsite meteorological measurements at one minute interval.

In this study, dispersion modeling was conducted based upon hourly measurement data for four periods with each period having consecutive five days from fall 2008 to summer 2009. Dates with rainfall, snowfall or other unusual conditions (e.g., instrument calibration dates) were excluded. The selection of the five days was set to be the periods with the least data points being excluded. In total, 435 valid hours were finally selected for the modeling practices.

2.2. AEROMOD Simulation

2.2.1 PM_{10} Emission Rate Determination

For tunnel-ventilated houses, the PM_{10} emission rate (ER) of house 4 was calculated based upon field measurements of PM_{10} concentration and house ventilation rate using Equation (1) [5] The ERs from other three tunnel houses (1 - 3) were assumed to be the same as it from the house 4 since these four houses were identical in building configurations, ventilation system controls, production management practices, and hen types/ages/populations

$$ER = C \cdot Q \tag{1}$$

where

ER = emission rate of the PM_{10}, μg s^{-1}

C = PM_{10} concentration at the source (ST1), µg m^{-3}

Q = ventilation flow rate of individual fan, m^3 s^{-1}

For cross-ventilated houses, since they shared the same production management practice with the tunnel-ventilated ones, also because they housed 2 additional row of caged laying hens as compared to the tunnel houses (8 rows *versus* 6 rows), the emission rate of cross-ventilated ones were assumed to be 8/6 times the emission rate of tunnel-ventilated houses.

For naturally ventilated houses, the ERs would be much less because of wet base manure management practice. Thus, it was assumed that the PM concentrations were 80% of the PM concentrations at the first floor of the tunnel ventilated houses for spring, summer and fall times. The ventilation of naturally ventilated houses was basically through the curtains on the side walls. The air flow velocity (V) was assumed to be the same as the wind speed vector in the direction perpendicular to the side walls of the naturally ventilated houses. The curtains were assumed to be fully opened for summer times, half opened for spring and fall times and fully closed in winter times. For winter times, the 20 ridge vents of the houses 102 & 103 were of the exhausts for air emissions. The house 93 did not have vents on the ridge, the emission from this house was insignificant in winter times, therefore, the ERs from houses 102 and 103 in winter times were computed by the following equation:

$$ER_{nat.} = f_1 \cdot f_2 \cdot f_3 \cdot ER_{tun.} \tag{2}$$

where

f_1 = factor due to wet base = 0.8

f_2 = factor due to ventilation = 0.4

f_3 = factor due to population = 70,000 hens in house 102 & 103/96,000 hens in house 4

ER_{tun} = total emission rate in tunnel ventilated house, ug s^{-1}

2.2.2. Source Inputs

Each fan in the mechanical ventilation systems was considered as an individual horizontal point source. The emissions from side wall curtains of the naturally ventilated houses in spring, summer and fall were considered as area sources; whereas in winter time the 20 ridge vents were considered as 20 capped point sources for houses 102 and 103. The stack heights and stack diameters of point sources were determined based upon the fan diameter and locations (**Table 2**).

Table 2. Stack heights and stack diameters for point sources.

Source description	Stack height (m)	Stack diameter (m)
Fans on the first floor of house 1 - 4	1.35	1.22
Fans on the second floor of house 1 - 4	3.84	1.22
Fans of house 5 - 6	1.35	1.22
Vents of house 102 & 103	5.5	1.54[a]

[a]Rectangular vents with an area of 1.86 m^2. A stack diameter of 1.54 m was used as it provides a same area.

The stack velocities for horizontal point sources were computed by Equation (3):

$$V = Q \cdot \frac{4}{\pi D_s^2} \tag{3}$$

where D_s = the stack (fan) diameter

Q = ventilation flow rate of individual fan, $m^3\ s^{-1}$

For naturally ventilated houses, the dimensions of the opening of the curtains were the dimension of the area source. For capped point sources, the stack velocities were assumed to be the same as wind velocities; for area sources, the exit velocities were assumed to be the wind velocities in the direction perpendicular to the side walls.

2.2.3. Plume-Rise Prediction

After completion of all the input files, hourly AERMOD simulations were conducted to derive plume rises for the selected data periods specified in **Table 3**. These hourly simulation followed the processes in the following diagram.

In **Figure 4**, the RD was compute by comparing the model predicted data (ΔC_p) with the field measurements (ΔC_m) using Equation (4):

Table 3. Data periods selected for modeling PM_{10} dispersion to derive plume rises through inverse AERMOD method.

Season	Year	Date period
Fall	2008	October 11-15
Winter	2008	February 4-8
Spring	2009	April 12-13 & April 16-18
Summer	2009	June 18-22

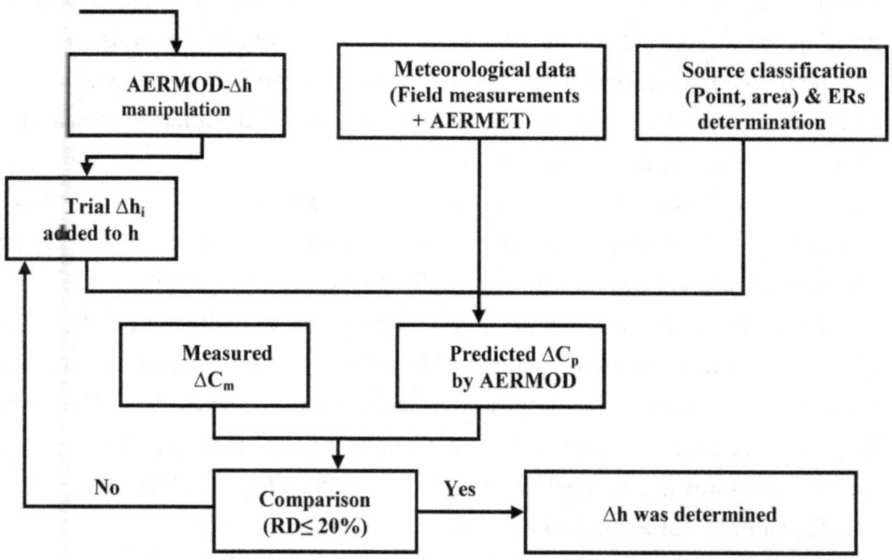

RD: relative difference between model production and field measurement; ΔC_m: measured PM_{10} concentration at the receptor (downwind minus upwind); ΔC_p: modeled PM_{10} concentration at the receptor; Δh: plume rise.

Figure 4. Flow chart of inverse-AERMOD modeling in predicting plume rises [52].

$$RD(\%) = \frac{\left|\Delta C_p - \Delta C_m\right|}{\Delta C_m} \tag{4}$$

2.3. Data Analysis

After all the hours were run and the Δhs were determined, seasonal, diurnal and weekly changes of plume rises were then analyzed. As the magnitude of plume rises may be affected by wind direction, wind speed and the temperature differences between ambient and the emission plume flow, these factors were chosen for multiple linear regression analysis. Also, the cloud cover and solar radiation were closely related to the stability class of the ambient environment, and were thus taken into consideration as influencing factors (predictors). Since momentum and thermal buoyancy had been the most important factors affecting plume rises, the total volume flow of the exhaust fans was then chosen to represent the momentum of the source. After selecting the possible factors affecting plume rises, a multiple linear regression analysis was conducted to identify the significance of each selected factor.

3. Results and Discussion

Out of the 435 hourly model runs, 87 valid plume rises were found. The reasons that the rest of the hours were not able to give out a plume rise value are (1) the ΔC_m for that hour was negative, meaning that the measured PM concentration of any possible downwind station was lower than the upwind station; (2) RD> 20%.

3.1. Seasonal Variations of the Plume Rise

Based on the prediction, all plume rises in each season were averaged and compared to see if the seasons would affect the magnitude of the plume rises (**Table 4**). The plume rises of spring and fall were higher than the other two seasons. In winter time, the average plume rise was much lower than other three seasons. This is probably because in winter times, the ventilation rate was not as high as other seasons. In addition, only 15 valid plume rises were obtained for winter time, among which only three hours were from daytime emissions, the plume rises may be affected by diurnal pattern. The average of these 15 plume rises may not well represent the seasonal mean. For summer time, as the thermal buoyancy was small due to small temperature differences between emission source and ambient air, the main plume rises were mainly due to the ventilation. However, despite the fact that summer had the highest ventilation rate, the plume rise in summer 2009 was not as much as those in fall 2008 and spring 2009. The plume rises for summer times need to be further investigated.

The standard errors in **Table 4** show that in spring and fall, the variation of plume rises is the highest. This is probably because of the high variation of house ventilation in response to the fluctuation of in-house temperature. Change of ventilation rate may cause the plume rise to change. On the other hand, in

Table 4. Mean plume rises by seasons.

Seasons	Fall 2008	Winter 2008	Spring 2009	Summer 2009
# of valid plume rises	20	15	33	19
Avg. plume rises (m)*	16.2[a]	7.9[b]	16.5[a]	14.3[c]
Standard error (m)	11.2	9.5	12.4	10.0
Avg. ventilation ($m^3\ s^{-1}$)	1421	324	1033	2277
Avg. ΔT (°C)	4.6	18.5	9.0	1.5

*Means with different letters are significant at 0.05.

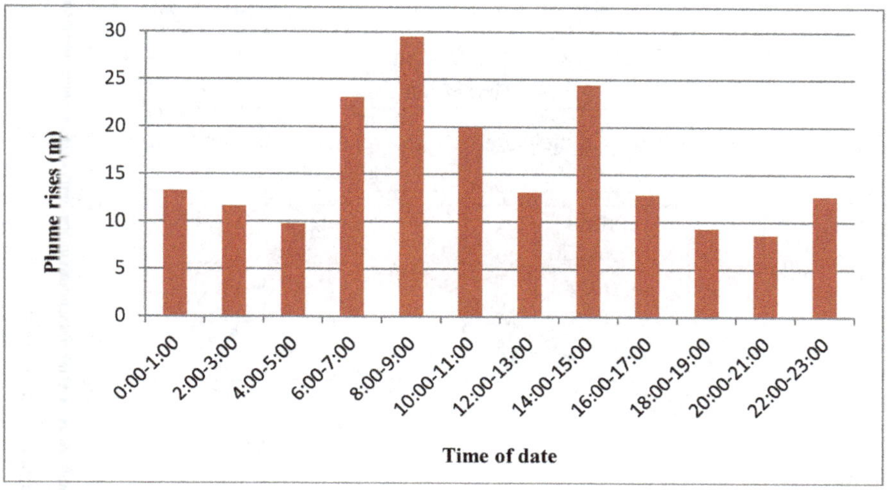

Figure 5. Diurnal variation of the plume rises.

summer and winter, the ventilation rate is relatively stable (either reached maximum ventilation in summer, or stayed at minimum ventilation rate in winter), leading to smaller standard errors for plume rises.

3.2. Diurnal Variation of the Plume Rise

To investigate diurnal variations of the plume rise, the plume rises and the time of a day were sorted together from earlier of the day till the latest ones. The plume rises were then grouped by each two hours from 0:00 to 23:00 and averaged within each group. In results, 12 mean plume rises were obtained and shown in **Figure 5**. As it can be seen, the lowest plume rise appeared at 2:00-3:00 or 18:00-21:00, the highest plume rise appeared at 8:00-9:00. There was no obvious diurnal impact on the plume rises by sorting the whole set of data. This was probably because that as the weather changed for different seasons, the ventilation varied, thus changing the diurnal patterns. Diurnal results seemed to indicate that the time when the peak plume rise appeared may be a matter of other parameters such as ventilation.

To analyze the response of the plume rises to the house ventilation, both ventilation rates and the plume rises for two representative dates were plotted in **Figure 6** & **Figure 7**. In general, the plume rises responded to the ventilation rates well. The higher the ventilation rate was, the higher the plume rise occurred.

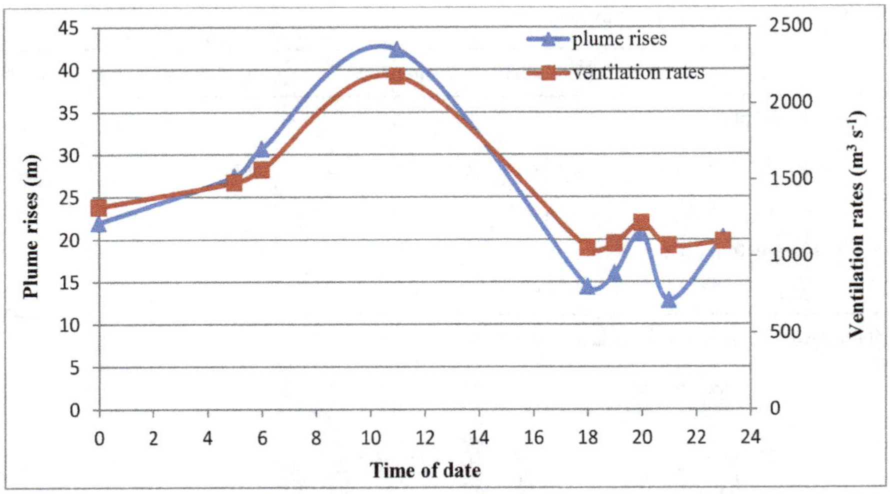

Figure 6. Diurnal variations of the plume rise and ventilation rate on October 11, 2008.

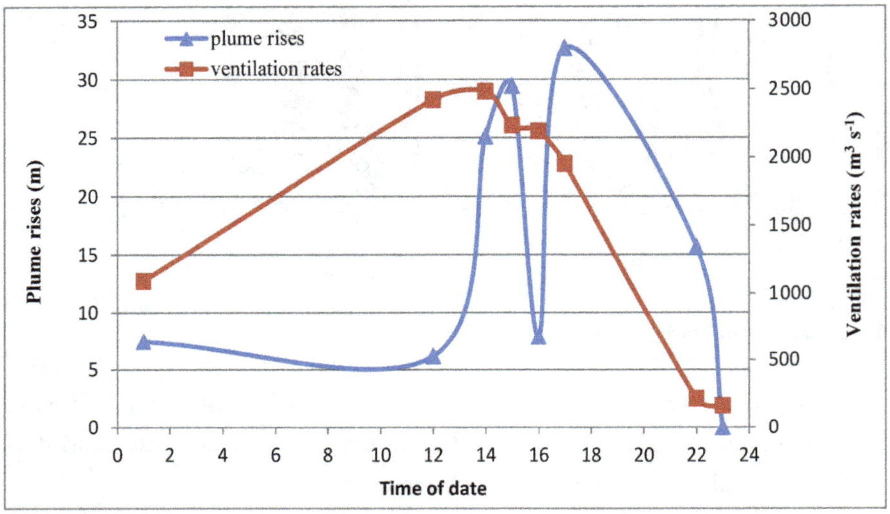

Figure 7. Diurnal variation of the plume rise and ventilation rate on April 16, 2009.

Pearson's correlation coefficients (ρ) tests were also applied for the two days. For October 11, 2008 (**Figure 6**), the relationship between ventilation and plume rises is quite firm ($\rho = 0.9766$) and the diurnal pattern of the plume rise followed very well with the diurnal pattern of the ventilation rate. For April 16, 2009, however, the relationship between the two was very poor ($\rho = 0.4391$). Although it seems high plume rises occurred at high ventilation rate, the plume rise was abnormally low at 16:00 on April 16, 2009 while the ventilation rate was not. A careful check of all possible related parameters at that hour was conducted for explanation. It was discovered that, except for the ranking of PM_{10} concentration levels among the four ambient stations, other factors (parameters) were not significantly different from the hourly data before and after this specific hour. At 15:00, 17:00 and even 14:00 of that day, the PM_{10} concentrations of the ambient stations all followed the order in ST4 > ST2 > ST5 > ST3; while at 16:00 the order was ST2 > ST4 > ST3 > ST5. It is unknown if this change of PM_{10} concentration ranking among station would have any impact.

3.3. Comparison of the Plume Rises on Weekday versus Weekends

Since the seasonal variation of the plume rises was observed, the comparison of plum rise on weekdays and weekends was conducted within one season. The average plume rises for weekdays and weekends are shown in **Table 5**.

At a significance level of 0.05, the plume rises during weekends were significantly higher than those during weekdays (p-value = 0.0019) in Fall. For spring, winter and summer, the difference between the two groups was not significant (p-value = 0.1551, 0.4779 and 0.8763, respectively). Comparison of the plume rises for Fall 2008 and Spring 2009 indicates no significant difference between the weekend plume rises for the two seasons, 19.76 m and 24.91 m; but there was a six times' difference for the weekday plume rises between these two seasons, 2.02 m and 12.78 m. The fact that Winter 2008 and Spring 2009 had the plume rise ratios of $\Delta h_{weekends}/\Delta h_{weekdays}$ in 2.02 and 1.95 implies the plume rises for weekends were about twice as much as the plume rises for weekdays.

For those hours providing valid plume rises, the predicted PM_{10} concentrations by AERMOD for ambient stations were observed to be smaller as the plume rise became larger. This means that if the ambient stations measured greater PM_{10} concentrations than the true contributions of the farm house emissions only, the inverse-AERMOD would produce smaller plume rise. This may explain the observation that the weekday plume rises tend to be smaller than weekend plume rise since there were some traffics around the farm, contributing PM_{10} to ambient stations. The PM_{10} contribution of the farm traffics were included in PM_{10} measurements at ambient stations and yet not included in emission rate for model simulation.

3.4. Multiple Linear Regression Analysis

To identify the significance of various factors (e.g., wind speed/direction, ambient/source temperature/RH, solar radiation, ventilation rate) that may affect the plume rise, the linear regression analysis was conducted and results of the analysis are listed in **Table 6**. The p-values of the multiple linear model shows that the wind speed had the strongest relationship with plume rise (p-value = 0.0134). Wind direction, ambient RH and total volume flow had a little relationship with the plume rises (p-value = 0.2285, 0.1709 and 0.2173, respectively). The possible reason for wind speed to be the most important factor would be the fact that the

Table 5. The impact of weekday and weekend on the plume rises.

Seasons	Average plume rises during weekdays (m)	Sample size	Standard error (m)	Average plume rises during weekends (m)	Sample size	Standard error (m)
Fall 08	2.02	4	2.02	19.76	16	9.53
Winter 08	6.23	11	9.79	12.60	4	7.93
Spring 09	12.78	23	10.20	24.91	10	13.39
Summer 09	14.10	14	10.80	14.90	5	8.30

Table 6. The multiple linear regression modeling results for the simulated plume rises.

	Estimate	Standard error	*p*-value
Intercept	426.6	192.4	0.0296
Wind direction	0.01474	−0.01214	0.2285
Wind speed	3.283	1.297	0.0134
Ambient temperature	297.2	789.2	0.7075
Source temperature	−298.6	789.3	0.7062
Temperature difference	296.5	789.1	0.7081
Ambient RH	0.1239	0.08902	0.1709
Cloud cover	0.2050	0.3617	0.5726
Solar radiation	−0.003875	0.01159	0.7391
Total volume flow	0.003245	0.02608	0.2173

exit velocities for naturally ventilated houses were estimated by the wind speeds, which strengthened the relationship between wind speeds and plume rises. It should also be noticed, the ground temperature could also be a possible factor affecting the plume rise of close to ground level emissions. However, no measurement of the ground temperature was done during the field data collection.

4. Summary and Conclusions

Based on the 87 predictions of plume rises of AERMOD simulations for the selected periods for four seasons, the following findings are obtained:

1) The mean plume rises were 16.2 m (SE = 11.2 m), 7.9 m (SE = 9.5 m), 16.5 m (SE = 12.4 m), 14.3 m (SE = 10.0 m) for fall, winter, spring and summer, respectively. More data are needed to support the prediction, especially for winter times.

2) The relationship between time of a day and the plume rises from animal housing systems did not demonstrate consistent diurnal patterns. However, in general, diurnal patterns of the plume rise followed the patterns of the ventilation rates.

3) The plume rises during weekends were significantly higher than that during weekends.

4) Wind speed had a very strong positive effect on the plume rises, whereas other parameters (*i.e.*, temperature, RH, ventilation, solar radiation and cloud cover) did not show significant impact (p-value > 0.1) on the magnitude of the plume rises.

Acknowledgements

This study was supported in part by NSF CAREER Award No. CBET-0954673 and USDA NIFA Higher Education Challenger Grant Program 2013-70003-20929. The PM source emission data and wind data were monitored under the National Air Emission Monitoring Study (NAEMS) Southeast Layer Operation, which was funded by the American Egg Board. Help from Qianfeng Li on field sampling is also acknowledged.

References

[1] Guo, L., Maghirang, R.G., Razote, E.B., Trabue, S.L. and McConnell, L.L. (2011) Concentrations of Particulate Matter Emitted from Large Cattle Feedlots in Kansas. *Journal of the Air & Waste Management Association*, **61**, 1026-1035. https://doi.org/10.1080/10473289.2011.599282

[2] Donham, K.J., Wing, S., Osterberg, D., Flora, J.L., Hodne, C., Thu, K.M. and Thorne, P.S. (2007) Community Health and Socioeconomic Issues Surrounding Concentrated Animal Feeding Operations. *Environmental Health Perspectives*, **115**, 317-320. https://doi.org/10.1289/ehp.8836

[3] Li, Q-F., Wang-Li, L., Shah, S.B., Jayanty, R.K.M. and Bloomfield, P. (2011) Fine Particulate Matter in a High-Rise Layer House and Its Vicinity. *Transaction of the ASABE*, **54**, 2299-2310. https://doi.org/10.13031/2013.40649

[4] Li, Q.-F., Wang-Li, L., Walker, J.T., Shah, S.B., Bloomfield, P. and Jayanty, R.K.M. (2012) Particulate Matter in the Vicinity of an Egg Production Facility: Concentrations, Statistical Distributions and Upwind and Downwind Comparison. *Transaction of the ASABE*, **55**, 1965-1973. https://doi.org/10.13031/2013.42359

[5] Li, Q-F., Wang-Li, L., Wang, K., Chai, L., Cortus, E.L., Kilic, I., Bogan, B.W., Ni, J-Q. and Heber, A.J. (2013a) National Air Emissions Monitoring Study's Southeast Layer Site: Part II. Particulate Matter. *Transaction of the ASABE*, **56**, 1173-1184.

[6] Li, Q.-F., Wang-Li, L., Jayanty, R.K.M., Shah, S.B. and Bloomfield, P. (2013) Ammonia Concentrations and Modeling Inorganic Particulate Matter in the Vicinity of an Egg Production Facility. *Journal of Environmental Science and Pollutant Research*, **21**, 4675-4685. https://doi.org/10.1007/s11356-013-2417-z

[7] Wang-Li, L., Li, Q. and Byfield, G.E. (2013) Identifications of Bioaerosols Released from an Egg Production Facility in Southeast U.S. *Journal of Environmental Engineering Science*, **30**, 2-10. https://doi.org/10.1089/ees.2011.0517

[8] Hu, D., Wang-Li, L., Simmons, O.D., Classen, J.J. and Osborne, J.A. (2015) Spatiotemporal Variations of Bioaerosols in the Vicinity of an Animal Feeding Operation Facility in the U.S. *Journal of Environmental Protection*, **6**, 614-627. https://doi.org/10.4236/jep.2015.66056

[9] Purdy, C.W., Clark, R.N. and Straus, D.C. (2007) Analysis of Aerosolized Particulates of Feedyards Located in the Southern High Plains of Texas. *Aerosol Science and Technology*, **41**, 497-509. https://doi.org/10.1080/02786820701225838

[10] NRC (2003) Air Emissions from Animal Feeding Operations: Current Knowledge, Future Needs. The National Academies Press, Washington DC.

[11] EPA (2004) Ohio's Largest Egg Producer Agrees to Dramatic Air Pollution Reductions from Three Giant Facilities. Environmental Protection Agency. http://yosemite.epa.gov/opa/admpress.nsf/b1ab9f485b098972852562e7004dc686/508199b8068c24a585256e43007e1230!OpenDocument

[12] EPA (2005) Air Quality Compliance Agreement for Animal Feeding Operations. https://www.epa.gov/afos-air/national-air-emissions-monitoring-study#afo-compliance-agreement

[13] Faulkner, W.B., Powell, J.J., Lange, J.M., Shaw, B.W., Lacey, R.E. and Parnell, C.B. (2007) Comparison of Dispersion Models for Ammonia Emissions from a Ground-Level Area Source. *Transactions of the ASABE*, **50**, 2189-2197.

https://doi.org/10.13031/2013.24094

[14] Hoff, S.J., Bundy, D.S. and Harmon, J.D. (2008) Modeling Receptor Odor Exposure from Swine Production Sources using CAM. *Applied Engineering in Agriculture*, **24**, 821. https://doi.org/10.13031/2013.25369

[15] Vieira de Melo, A.M., Santos, J.M., Mavroidis, I. and Reis Junior, N.C. (2012) Modeling of Odour Dispersion around a Pig farm Building Complex Using AERMOD and CALPUFF. Comparison with Wind Tunnel Results. *Building Environment*, **56**, 8-20.

[16] Ritter, W.F. and Rao Chitikela, S. (2007) Modeling Ammonia and Odor Emissions from Livestock and Poultry Facilities: A Review. World Environmental and Water Resources Congress 2007: Restoring Our Natural Habitat. https://doi.org/10.1061/40927(243)147

[17] Potter, R.A. (2014) Understanding, Characterizing and Modeling Ammonia Dispersion near CAFOs. M.S. Thesis, Department of Marine, Earth and Atmospheric Sciences, North Carolina State University, Raleigh.

[18] Zhu, J., Jacobson, L.D., Schmidt, D.R. and Nicolai, R. (2000) Evaluation of INPUFF-2 Model for Predicting Downwind Odors from Animal Production Facilities. *Applied Engineering in Agriculture*, **16**, 159-164. https://doi.org/10.13031/2013.5068

[19] Sarr, J.H., Goïta, K. and Desmarais, C. (2010) Analysis of Air Pollution from Swine Production by using Air Dispersion Model and GIS in Quebec. *Journal of Environmental Quality*, **39**, 1975-1983. https://doi.org/10.2134/jeq2009.0199

[20] Bonifacio, H.F., Maghirang, R.G., Auvermann, B.W., Razote, E.B., Murphy, J.P. and Harner III, J.P. (2012) Particulate Matter Emission Rates from Beef Cattle Feedlots in Kansas-Reverse Dispersion Modeling. *Journal of the Air & Waste Management Association*, **62**, 350-361. https://doi.org/10.1080/10473289.2011.651557

[21] Bonifacio, H.F., Maghirang, R.G., Razote, E.B., Trabue, S.L. and Prueger, J.H. (2013) Comparison of AERMOD and WindTrax Dispersion Models in Determining PM_{10} Emission Rates from a Beef Cattle Feedlot. *Journal of the Air & Waste Management Association*, **63**, 545-556. https://doi.org/10.1080/10962247.2013.768311

[22] O'Shaughnessy, P.T. and Altmaier, R. (2011) Use of AERMOD to Determine a Hydrogen Sulfide Emission Factor for Swine Operations by Inverse Modeling. *Atmospheric Environment*, **45**, 4617-4625.

[23] Flesch, T.K., Wilson, J.D., Harper, L.A. and Crenna, B.P. (2005) Estimating Gas Emissions from a Farm with an Inverse-Dispersion Technique. *Atmospheric Environment*, **39**, 4863-4874.

[24] Flesch, T.K., Wilson, J.D., Harper, L.A., Todd, R.W. and Cole, N.A. (2007) Determining Ammonia Emissions from a Cattle Feedlot with an Inverse Dispersion Technique. *Agricultural and Forest Meteorology*, **144**, 139-155.

[25] Flesch, T.K., Harper, L.A., Powell, J.M. and Wilson, J.D. (2009) Inverse-Dispersion Calculation of Ammonia Emissions from Wisconsin Dairy Farms. *Transactions of the ASABE*, **52**, 253-265. https://doi.org/10.13031/2013.25946

[26] Jahne, M., Rogers, S., Lopez, K., Kim, S., Grimberg, S. and Holsen, T. (2013) Inverse Dispersion Modeling of Bioaerosol Emissions from Manure Application Sites. *ASABE Annual International Meeting*, Kansas City, 21-24 July 2013, Paper No. 131594700.

[27] Seinfeld, J.H. and Pandis, S.N. (1997) Atmospheric Chemistry and Physics: From Air Pollution to Climate Change. Wiley Interscience.

[28] Turner, D.B. (1994) Workbook of Atmospheric Dispersion Estimates: An Introduction to Dispersion Modeling. 2nd Edition, Lewis Publishers.

[29] Yadigaroglu, G. and Munera, H.A. (1987) Transport of Pollutants—Summary Review of Physical Dispersion Models. *Nuclear Technology*, **77**, 125-149. https://doi.org/10.13182/NT87-A33979

[30] Crane, G., Panofsky, H. and Zeman, O. (1977) A Model for Dispersion from Area Sources in Convective Turbulence. *Atmospheric Environment*, **11**, 893-900.

[31] Carney, P.G. and Doddt, V.A. (1989) Comparison between Predicted and Measured Values for the Dispersion of Malodors from Slurry. *Journal of Agricultural Engineering Research*, **44**, 67-76.

[32] Briggs, G.A. (1969) Mathematical Analysis of Chimney Plume Rise and Dispersion: Optimum Formulas for Buoyant Plume Rise. *Philosophical Transactions of the Royal Society of London*, **265**, 197-203. https://doi.org/10.1098/rsta.1969.0048

[33] Cooper, C.D. and Alley, F.C. (2002) Air Pollution Control: A Design Approach. Waveland Press, Prospect Heights.

[34] Weber, A.H. (1976) Atmospheric Dispersion Parameters in Gaussian Modeling. EPA, Washington DC, EPA-600/4-76-030A.

[35] Bennett, M., Sutton, S. and Gardiner, D.R.C. (1991) Measurement of Wind Speed and Plume Rise with a Rapid-Scanning LIDAR. *Atmospheric Environment*, **26A**, 1675-1688.

[36] Bennett, M., Sutton, S. and Gardiner, D.R.C. (1992) An Analysis of LIDAR Measurement of Buoyant Plume Rise and Dispersion at Five Power Stations. *Atmospheric Environment*, **26A**, 3249-3263.

[37] Contini, D. and Robins, A. (2001) Water Tank Measurements of Buoyant Plume Rise and Structure in Neutral Crossflows. *Atmospheric Environment*, **35**, 6105-6115.

[38] Michioka, T., Sato, A., Kanzaki, T. and Sada, K. (2007) Wind Tunnel Experiment for Predicting a Visible Plume Region from a Wet Cooling Tower. *Journal of Wind Engineering and Industrial Aerodynamics*, **96**, 741-754.

[39] Hanna, S.R., Briggs, G.A. and Chang, J.C. (1997) Lift-Off of Ground-Based Buoyant Plumes. *Journal of Hazardous Material*, **59**, 123-130.

[40] Meroney, R.N. (1979) Lift Off of Buoyant Gas Initially on the Ground. *Journal of Industrial Aerodynamics*, **5**, 1-11.

[41] Briggs, G.A., Britter, R.E., Hanna, S.R., Havens, J.A., Robins, A.G. and Snyder, W.H. (2001) Dense Gas Vertical Dilution over Rough Surfaces: Results of Wind Tunnel Studies. *Atmospheric Environment*, **35**, 2265-2284.

[42] Pasquill, F. (1961) The Estimation of the Dispersion of Windborne Material. *Meteorology Magazine*, **90**, 33-40.

[43] Briggs, G.A. (1971) Plume Rise: A Recent Critical Review. *Nuclear Safety*, **12**, 15-24.

[44] Briggs, G.A. (1984) Plume Rise and Buoyance Effects. In: Randerson, D., Ed., *Atmospheric Science and Power Production*, U.S. Department of Energy, 327-366.

[45] Calder, K.L. (1977) Multiple-Source Models of Urban Air Pollution—Their General Structure. *Atmospheric Environment*, **11**, 403-414.

[46] Pasquill, F. (1971) Atmospheric Diffusion of Pollution. *Quarterly Journal of the Royal Meteorological Society*, **97**, 369-395. https://doi.org/10.1002/qj.49709741402

[47] Schatzmann, M. (1979) An Integral Model of Plume Rise. *Atmospheric Environment*, **13**, 721-731.

[48] EPA (2003) AERMOD: Latest Features and Evaluation Results. EPA-454/R-03-003. http://www.epa.gov/ttn/scram/7thconf/aermod/aermod_mep.pdf

[49] Weil, J.C., Corio, L.A. and Brower, R.P. (1997) A PDF Dispersion Model for Buoyant Plumes in the Convective Boundary Layer. *Journal of Applied Meteorology and Climatology*, **36**, 982-1003. https://doi.org/10.1175/1520-0450(1997)036<0982:APDMFB>2.0.CO;2

[50] Weil, J.C. (1985) Updating Applied Diffusion Models. *Journal of Applied Meteorology and Climatology*, **24**, 1111-1130. https://doi.org/10.1175/1520-0450(1985)024<1111:UADM>2.0.CO;2

[51] Wang, L., Li, Q.-F., Wang, K., Bogan, B.W., Ni, J.-Q., Cortus, E.L. and Heber, A.J. (2013) National Air Emissions Monitoring Study's Southeast Layer Site: Part I. Site Characteristics and Monitoring Methodology. *Transaction of the ASABE*, **56**, 1157-1171.

[52] Ying, M. (2012) Modeling Plume Rise of Air Emissions from Animal Housing Systems: Applications of CFD and AERMOD. M.S. Thesis, Department of Biological & Agricultural Engineering, North Carolina State University, Raleigh.

10

Waste-to-Hope: Measuring Sustainability Benefits of Product Philanthropy Partnership

Oral S. Saulters*, Brijesh Krishnan, Mary Beth Harris, Hani Bashour, Laura Bellows, Evan J. Ringquist, Diane Henshel

School of Public and Environmental Affairs, Indiana University, Bloomington, IN, USA
Email: *osaulter@indiana.edu

Abstract

This study investigates the sustainability-related impacts of the Framing Hope program, an innovative private and nonprofit sector partnership between the nonprofit Good360 and The Home Depot. Specifically, this paper estimates the energy savings, landfill space not filled, and the energy- and landfill-related cost savings associated with the Good360 product philanthropy program with The Home Depot, illustrating that benefits accrue to both company and communities by linkages between environmental, socioeconomic, and energy dimensions. Findings suggest that by redirecting products from the waste stream into usable community resources, considerable landfill space, costs, and energy savings were realized. Product donations can serve as an important platform for sustainable community development and capacity building. This research adds to the growing body of knowledge on environmental performance, corporate giving, and cross-sector partnerships in sustainable and social entrepreneurship.

Keywords

Sustainability, Product Philanthropy, Corporate Social Responsibility

1. Introduction

Sustainable development and sustainability-related initiatives have become increasingly important considerations for communities, organizations, and cross-sector collaborations [1] [2] [3] [4]. Many of these efforts involve partnerships between private and nonprofit organizations [5] [6]. Beyond intuitive and anecdotal notions of "win-win" benefits, there is a growing body of research assessing the performance outcomes of these corporate/nonprofit collaborations for nonprofits and their clients, increasingly within the context of social entrepreneurship [7]-[15]. Similarly, research and community outreach addressing social justice and equity within ecological limits, for all stakeholders, is being integrated into "just sustainabilities" decision making perspectives [16].

This study examines the impact of the Framing Hope Product Donation program for nonprofit organizations on local community sustainability. Initial work was conducted in a pilot phase [17] followed by subsequent analysis. As one of the first investigations of this type, this research specifically evaluates environmental and energy dimensions of product philanthropy. To begin unpacking the multidimensional effects of the Framing Hope program on community sustainability, three research questions were addressed: What are the landfill volume and tipping cost savings? What are the manufacturing energy savings? What are the energy consumption savings?

2. Product Philanthropy and Corporate Social Responsibility

Corporate philanthropy has a long and controversial history including involvement of American industry titans such as Andrew Carnegie, Henry Ford, and J.D. Rockefeller. For example, Carnegie stressed a duty to the most beneficial results for the community when the more common neoclassical approach emphasized strict obligation to shareholders [18] [19] [20]. Corporate philanthropy is driven by various motives including instrumental reasons for profit maximization, political legitimacy, and altruistic efforts for social responsibility [21] [22] [23]. Today Carnegie's perspective is labeled "Corporate Social Responsibility" (CSR). In its early modern incarnation, CSR was described as "the obligations of businessmen to pursue those policies, to make those decisions, or to follow those lines of action which are desirable in terms of the objectives and values of our society" [24]. More recently, Carroll has described CSR as a pyramid: the foundation is economically based (*i.e.*, be profitable); followed by legal requirements (obey the law); ethical responsibilities (do no harm); and ultimately at the apex, the philanthropic responsibility to contribute resources to the community [25]. In recent years, companies of all sizes and classifications have been implementing CSR and related initiatives with differing motivations, approaches, and outcomes [26]-[32]. In particular, sustainability and shared value programs have been increasing; many of these programs include substantial charitable donations and intentional partnerships with nonprofit organizations [33] [34] [35] [36] [37]. With more than 1.5 million nonprofit organizations in the US, there are tremendous opportunities for robust collaborations of this sort [38] [39] [40].

2.1. Product Philanthropy as an Element of Corporate Social Responsibility

Philanthropic giving is an important component of most CSR strategies. By one estimate, over the past forty years corporate giving has grown from $0.85 billion in 1971 to $14.56 billion in 2010 [41] (see **Figure 1**).

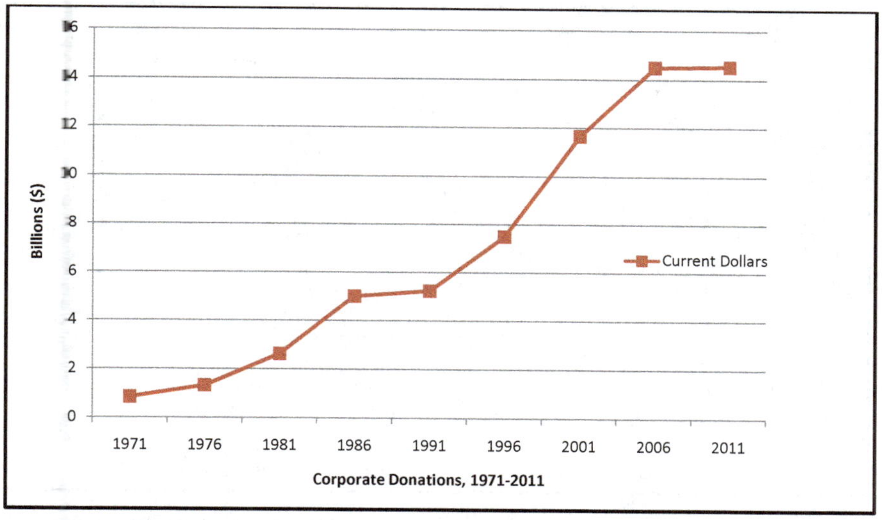

Figure 1. Estimated Trend in Corporate Giving, 1971 - 2011 (Adapted from Giving USA 2012).

A sizeable proportion of philanthropic donations are in the form of monetary contributions from corporations and grants and gifts made by corporate foundations. At the same time, a majority of these donations take the form of non-cash in-kind contributions made through corporate giving programs. In-kind contributions characterize direct or indirect non-cash donations of pro-bono services or products assessed at fair market value [41].

Corporate product donations represent a subset of all in-kind giving, distinct from any other corporate gifts of time and expertise. Product donations involve many diverse goods such as food, pharmaceutical and health care products, computer equipment, and other retail products [42]. Product donation programs are becoming an increasingly popular component of corporate management systems. As shown in **Figure 2**, pharmaceutical companies have been the largest contributors of in-kind giving in recent years, followed by computer equipment and information technology service companies. In fact, eight of the top ten corporate donors come from these two industries [43].

Product donations by private businesses represent an important source of support for nonprofits and have contributed meaningfully to the expansion of programs in national nonprofit organizations [44]. Despite the importance of corporate product donations, the value of in-kind donations has not been consistently measured over time [41]. Among the top 20 corporate donors in **Figure 2**, the value of product donations equaled nearly $12 billion, and product donations accounted for 81.5 percent of all corporate giving. This degree of corporate product philanthropy may not be typical, however. Data from 110 large companies show that while corporate giving increased 23 percent between 2007 and 2010, product donations increased at a much faster rate and accounted for approximately 60 percent of corporate giving during this period [45] see **Figure 3**).

Four factors help account for the recent increase in product donations: 1) Firms are replacing cash donations with product philanthropy during highly

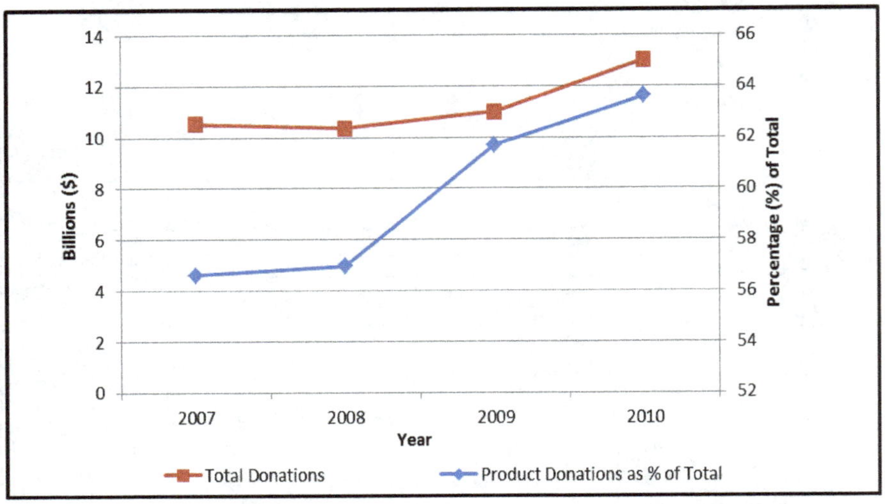

Figure 2. Survey of Corporate Giving and Product Donations (Adapted from CECP, 2011).

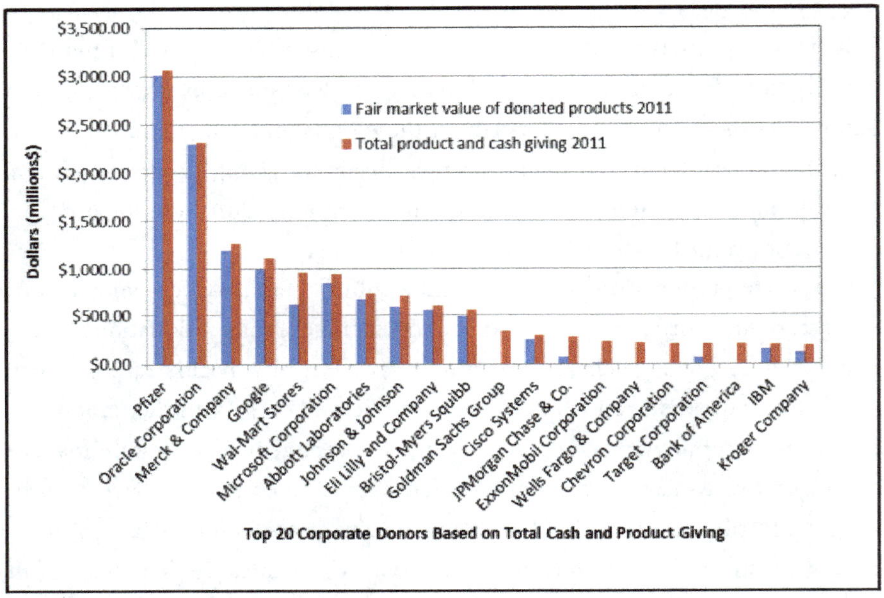

Figure 3. Top 20 Corporate Donors Based on Total Cash and Product Donations (Adapted from Chronicle of Philanthropy, 2012).

uncertain economic conditions that tend to generate excess inventory and lower cash balances; 2) Product donations are more visible in the long-term; 3) Product donation is a more environmentally sustainable option over landfill disposal; and 4) companies can capture financial advantages from the special tax treatment of product donations [46] (**Figure 4**).

2.2. An Example of Corporate Product Philanthropy: Framing Hope

Started in 2008, Framing Hope is a collaboration between The Home Depot, The Home Depot Foundation, Good360 (the nation's largest nonprofit specializing in product donations), and local nonprofit organizations. In 2010, Good360

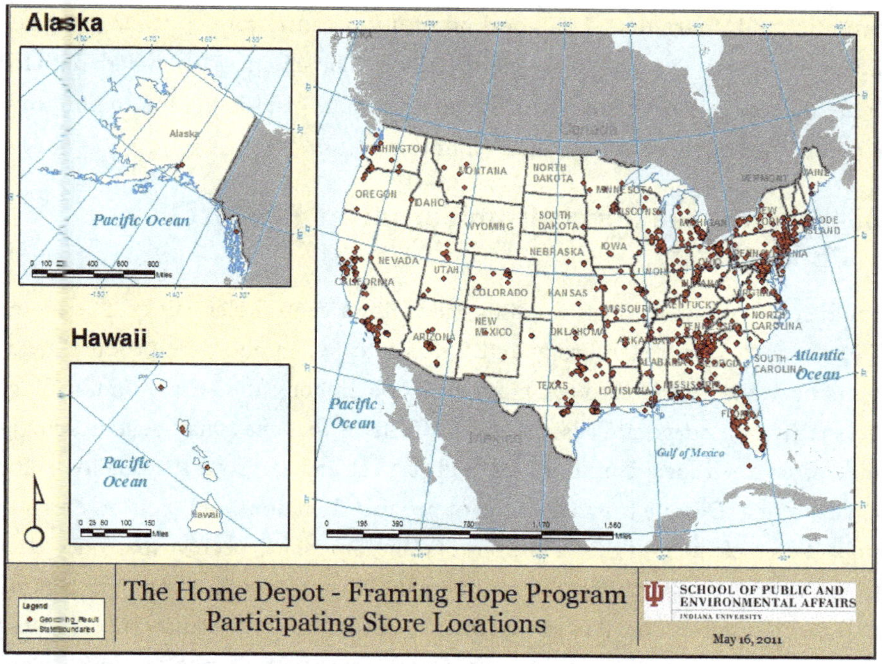

Figure 4. Framing Hope Program Locations in the United States.

handled more than $350 million in product donations from dozens of corporate partners. Through Framing Hope, The Home Depot stores are matched with nonprofits in their given communities. The nonprofits must be 501(c) (3) organizations and are required to go through an application and screening process to determine their eligibility. The nonprofit organizations must agree to use the products for charitable purposes and not sell, trade, barter, or auction any of the donations in accordance with Federal Tax Code 170(e)3, using the products only to help the poor, the sick, and children. The nonprofit organizations must provide support in housing/homelessness, youth/senior housing, civic/cultural issues, community re-building, revitalization, renovation or other similar projects [47]. In engaging in the partnership, the nonprofits must be willing to accept the products donated by the stores on a weekly basis. Depending on the particular store's retail activities and inventory, the products can range widely from week to week. One week the items may include bathroom fixtures, hammers, and an assortment of rugs; the following week, products may be doors, windows, and lumber. Guidance information and training are offered to help nonprofit organizations make use of and share items.

Based on initial success, the program was expanded in 2009 to offer more opportunities for smaller nonprofit organizations to participate that might not have the staff and infrastructure capacity to support the weekly pickups, storage, and use of all donated products. Therefore, the program launched the Framing Hope Warehouse Strategy in larger communities with higher volume to connect The Home Depot stores, large nonprofit organizations with the logistical capability to redistribute the products, and smaller nonprofit organizations. By the end of 2012, Good360 had matched The Home Depot stores in over 1000 communities with more than 1200 local nonprofit organizations to receive product donations [47]. Through these partnerships, The Home Depot had donated more than 10 million items with a fair market value of $150 million to nonprofit social service organizations in local communities.

3. Assessing the Effects from Corporate Product Philanthropy

Product donations through Framing Hope are a central element of The Home Depot's corporate philanthropy and CSR strategies. In addition, product donations like those made through Framing Hope embody the "triple bottom line" perspective on corporate sustainability efforts in that they may benefit companies, strengthen local communities, and protect and improve the quality of the environment. Despite the magnitude of product donations and their potential to impact each element of the "triple bottom line," there has been virtually no effort to evaluate the sustainability-related benefits of product donations for companies, communities, or the environment. Furthermore, because the Framing Hope program represents one company partnering with many nonprofit organizations, it offers a uniquely specific and central source of data that enable a quantitative evaluation of these sustainability-related benefits of the program.

3.1. Effects for Companies

The analysis of the business case for the Framing Hope program suggests that in-kind philanthropy is a prudent process that can benefit both the corporation and civil society including recipient nonprofit organizations [46]. Product donations are generally more financially beneficial than cash donations because of the special IRS tax deduction pursuant to Tax Code δ170(e)(3). Beyond financial grounds, there are other advantages to product donations over cash donations such as targeted visibility because of the direct pathway to end users rather than being used to cover administrative expenses by nonprofits, a concern evaluated by charity-rating groups like Charity Watch [43]. Other benefits are derived from the relative transaction simplicity for both companies and their nonprofit partners.

Strategic product donation programs can enhance corporate image by "signaling" to stakeholders and complementing advertising and marketing efforts. With a strong public image, employees can demonstrate stronger commitment and trust [33] [48]. Product donation programs can attract new talent, customers, and investors as consumers are demanding more product philanthropy and social purpose [49]. Moreover, these ventures can support a business friendly environment by establishing legitimacy with stakeholders including regulatory agencies and local governments which tend to grant greater latitude to firms with such socially responsible programs [46] [50]. Another set of key relationships in product donation collaborations are those with intermediary nonprofits such as Good360 and the National Association for the Exchange of Industrial Resources (NAEIR). Intermediaries, with their extensive logistical and charity networks, can provide cost advantages through assistance in matching, lowering administrative costs, simplifying logistics, ensuring brand security, and gaining positive media coverage. Generally, it is recommended that corporations research individual intermediaries carefully before engaging in partnerships to explore potential organizational compatibility and mission fit.

3.2. Effects for Communities

A survey and follow-up evaluation of the Framing Hope effects on organizational partners estimated that over 450,000 households benefitted from the program with a multimillion dollar value of received products [34]. The study found that over 60% of nonprofits share the donated products with other organizations. Challenges were noted with participating charity capacity such as excess inventory, transportation, storage, and lack of performance measures. Many nonprofits were able to leverage product donations to start new programs, expand services, and obtain additional funding by using the products as a match for grant requirements. Another key finding included in Gazley and Abner [33] (relates to the need for improving corporate communications of program objectives with local stores and improving training of employees to support those goals. Local context matters, as Paarlberg and Meinhold [51] have found with the United Way's nationwide Community Impact program; there can be significant variation in implementation and allocation practices at the community level driven by local logics. Further research on the complex nexus between organizational capacity and performance outcomes in the context of product philanthropy remains an important area of investigation.

According to the Consumer Expenditure Survey [52], people in lower economic groups spend a significantly higher percentage of their income on expenditure categories such as energy and utilities, housekeeping supplies, and household furnishings, areas directly related to products donated through the Framing Hope Program. Even small corporate donations can contribute substantively to household budgets. With almost no information in the research literature on the effects of product donations on meeting the needs of families and individual end users, Seefeldt and Abner [52] conducted a pilot case study to gather baseline data. The researchers noted the complexities in identifying, coordinating, and conducting interviews of Framing Hope end users. Their brief analysis addressed recipients as both organizations and individuals. For the nonprofits, donated products facilitated less dependence on financial donations and streamlined completion of service projects. For low-income individual end users, product philanthropy fostered self-sufficiency, provided higher quality products than thrift stores, and enhanced community life through empowerment for home improvements and social interactions at the neighborhood level [52]. This research suggests that while most scholarly efforts have focused on top down processes and estimated effects, empirical studies of actual stakeholders are needed to improve metrics of program impacts and performance dimensions.

3.3. Effects on the Environment

In evaluating the impacts from the sustainability triple bottom line, the final perspective, environmental outcomes, remained unknown. This paper begins to address the data gap. The remainder of this paper presents the three approaches used to quantify the sustainability-related benefits of the Framing Hope program, focusing particularly on environmental and social benefits. The two environmental benefits considered include landfill space preserved by not putting the reused products in landfills and the energy saved both in manufacturing the reused materials and energy saved by the receiving community, assuming that these products either replaced less energy efficient models or enabled behavioral energy savings. Cost savings associated with reduced landfill tipping fees and the cost of energy produced are also quantified. The paper concludes with a discussion of the results, placing this analysis within the context of the social entrepreneurship and social capitalism literature that uses a Blended Value approach to quantifying the social impacts of philanthropy and social entrepreneurship.

4. Measuring the Environmental Benefits of Framing Hope

4.1. Research Goals and Guiding Questions

The goal for this work was to assess the environmental and related social benefits gained by removing the donated Home Depot products from the waste stream and facilitating their reuse by donating them to nonprofits (for community redistribution) through the Framing Hope program.

Three questions guided this research: 1) How much volume is saved in landfills and what is the cost saved? 2) What is the embodied energy saved for products not manufactured? 3) How much energy are consumers saving by using the donated products that are designated energy efficient?

4.2. Data Sources, Verification, and Processing

Data were provided by Good360 and The Home Depot. The Home Depot has the SKU (stock keeping unit as unique identifier) of each donated product and through its Framing Hope web portal, allows integrated tracking, reporting, and verification functions to be performed. Data on donated products from February 1, 2008 through May 31, 2010 were evaluated. During this period, nearly 4.7 million products were donated representing 57,341 SKUs, worth approximately $64.7 million. Additionally, The Home Depot provided information on the dimension and weight of some products, although only 51.2 percent were reported in the data. The Home Depot also provided information on the make and model for certain products with energy usage value; however, only not all products had this information available. Finally, The Home Depot also provided their three tiered classification system for SKUs to assist with identifying products. After merging the two datasets, The Home Depot datasets were checked to ensure that there were no SKUs with inconsistent information (for example, lumber that was donated in different lengths).

4.3. Quantifying the Environmental Benefits of Donated Products

In 2010, Americans generated nearly 250 million tons of municipal solid waste with roughly 35 to 45 percent coming from commercial and institutional sources [53]. Approximately 136 million tons of waste (54.2 percent of total) was disposed of in landfills in 2010, down from nearly 89 percent of waste generated in 1980. Before the Framing Hope program was established in 2008, most of the returned, unsold, displayed, overstocked, and discontinued products ended up in local landfills. Many of the products donated through Framing Hope are building supplies and materials, plumbing products, assorted household and cleaning items. These types of products are often used for construction and renovation activities, and result in construction and demolition (C & D) debris—a type of solid waste that tends to be larger and is particularly disparate in form, physical characteristics, and materials [54]. Construction and demolition debris is state rather than federally regulated. The state programs vary widely in terms of legal definitions and management requirements [55] [56]. Organizations such as the U.S. Agency for Toxic Substances and Disease Registry (ATSDR) and federal, state, and local partners have noted community health and worker safety concerns with C&D landfills including those derived from the generation of hydrogen sulfide emissions associated with the decomposition of gypsum drywall and organic debris [57] [58] [59].

The nearly 14 tons of items donated through the Framing Hope program represent an important opportunity for product philanthropy. By shifting these products from waste streams into useful product resources, community sustainability can be improved in multiple ways. As an initial quantitative estimation of the sustainability impact of diverting these products, we quantified preserved landfill space (a community benefit), the cost savings associated with not putting these products in landfills (a business benefit), the energy embodied in these products through the manufacturing process (a global benefit), and the energy saved as well as the cost of energy saved by the receiving community members who were given energy efficient (USEPA Energy Star program) appliances and other materials that either replaced less energy efficient appliances or products or encouraged energy saving behaviors (e.g., using a fan; a community benefit).

1) Question #1a: How Much Volume Is Saved in Landfills by Removing the Donated Products from the Waste Stream?

Like many retail stores, The Home Depot disposed of much of its excess inventory as solid waste. This practice contributed to the disposal of new and useable products, consumption of valuable landfill space, and resulted in financial expenditures for waste management, fees, and future liability. The Framing Hope program provides a distinct, more sustainable alternative to disposal of these unwanted products.

The Home Depot database included data on relevant product characteristics including quantity, weight, and dimensions. Complete product information including dimensions, needed to estimate volume, and weight were available for only about half of the donated products. To fill in the gaps, the 57,341 products were sorted into 23 categories based on key material types (see **Table 1**). Within each of the individual categories, based on known weights, volumes, and quantities, average values and total values for the categories were calculated. Missing dimensions and weight values were estimated based on the averages for similar products.

Four assumptions were made about the data. 1) All products were indeed diverted from landfills. 2) Product SKUs were unique identifiers for products. 3) The product categories accurately classify individual items. 4) The reference data for each category was representative enough to fill in missing values of weight and volume. Using these assumptions, the total weight for the donated products was estimated to be **13,913 tons.**

To calculate the total landfill volume of donations, the actual expected landfill volume had to be estimated from the volume calculated from the product dimensions. Most products are compacted before being put into landfills [60] [61]

Table 1. Product Categories and Number of Stock Keeping Units (SKUs).

Category #	Category Name	Number of SKUs
1	Additives & Adhesions	209
2	Batteries	90
3	Cement Based Products	252
4	Ceramic Materials	2127
5	Construction Chemicals	421
6	Construction Materials	451
7	Construction Natural Materials	517
8	Electrical & Electronic Products	9705
9	Electrical Cables & Wires	654
10	Fiberglass	424
11	Glass Products	667
12	Metal Plastic Products	3464
13	Metal Plastic Wood Products	1498
14	Metal Products	18,552
15	Metal Rubber Products	305
16	Natural Products	2622
17	Organic Products	1498
18	Paints	62
19	Paper Cardboard Products	282
20	Plastic Products	6402
21	Rubber Products	562
22	Wood Furniture	2202
23	Wood Products	4375

[62]. To determine the landfill volume saved, a final post-compression landfill volume had to be calculated, based on the ratio between the pre-compaction density (157 lbs/yd^3) and the target post-compaction density (700 lbs/yd^3; a value provided by a Republic Services representative in Indiana). The estimated compression ratio used (1:4.456) was then applied to the initial volume using the following equation:

*Total Volume Reduced = Total Volume of Products * Compression Ratio* (*Equation* (1))

The total initial volume for the donated products was estimated to be **4,782,320 ft^3**. The estimated reduction in volume of landfill space was determined to be **1,073,271 ft^3** or **39,751 yd^3**.

2) *Question #1b: What Is the Cost Saved By Not Putting These Products in Landfills?* Tipping costs are dependent on weight. To arrive at total tipping costs saved by Framing Hope, the total product weight was multiplied by the per-ton tipping cost ($43.99 per ton, reported by Republic Services) per the following equation:

*Total Cost Savings = Total Weight of Products * Tipping Fee* (*Equation* (2))

Using the average tipping fee of $43.99/ton, and the total calculated weight of 13,913 tons, the total tipping fee cost savings was **$612,023**.

A second calculation method used to enhance the accuracy of our results integrated more micro-level information, including geospatial data. Specifically, the physical location of individual Home Depot stores and their community nonprofit partners that participated in the program were used to estimate location specific cost savings. Hierarchical aggregation was employed for the donated product information (weight and volume data) at each store location and across each state. The actual tipping fees (averaged across surveyed locations) for each state were identified [63] and used to calculate total cost savings at the state level. In addition, a more current typical average compaction target density of 1200 lbs/yd^3 (used for C&D waste and roughly 70% higher than the initial density identified by Republic Services in Indiana) was used [64] to calculate the landfill volume effectively offset. **Table 2** summarizes the state level results. These yielded overall cost savings of **$559,645** and volume savings of **626,075 ft^3** or **21,188 yd^3**.

3) Question #2: What Is the Energy Saved for Products Reused?

Recovering Energy Embodied in Product Donations

Another important measure of the impact of the Framing Hope program involves accounting for the total sum of lifecycle energy flows required to produce the donated products. When products are used and consumed, on a fundamental level energy is being transformed and used. Energy analysis is an attempt to estimate how much energy is required to provide goods and services. A key aspect of energy analysis is determining the embodied energy which is the total (direct and indirect) energy required for the production of economic or environmental goods and services [64].

Table 2. Tipping Fees, Weights, and Volumes of Donated Products for Each State.

State	Tipping Fee ($/Ton)	Weight (lbs)	Volume (in^3)	Total Cost ($)
AK	43.83	4.90E+04	1.28E+07	1073
AL	25	3.28E+05	9.60E+07	4095
AR	35	1.35E+05	3.92E+07	2360
AZ	29.24	4.75E+05	1.41E+08	6944
CA	43.83	2.34E+06	8.81E+08	51331
CO	30.47	1.87E+05	5.28E+07	2851
CT	63	1.37E+05	4.01E+07	4328
DC	55.45	2.61E+04	8.10E+06	724
DE	58.9	4.48E+04	9.72E+06	1319
FL	37	2.66E+06	6.86E+08	49,139
GA	34.92	2.74E+06	8.37E+08	47,771
HI	43.83	8.48E+04	1.67E+07	1859
IA	40.71	5.24E+04	1.23E+07	1066
ID	46.16	1.20E+04	3.78E+06	278
IL	35.5	8.23E+05	2.56E+08	14,606
IN	29.57	5.06E+05	1.50E+08	7476
KS	30	1.76E+05	5.00E+07	2640
KY	29.21	3.17E+05	8.25E+07	4632
LA	46	2.29E+05	6.16E+07	5278
MA	72	4.76E+05	1.23E+08	17, 144
MD	52	1.73E+06	5.48E+08	45,051
ME	60	6.30E+04	1.81E+07	1891
MI	34.7	1.10E+06	2.99E+08	19, 121
MN	50	2.30E+05	6.38E+07	5756
MO	33.73	2.93E+05	8.51E+07	4942
MS	25	9.94E+04	3.45E+07	1242
MT	42	3.04E+04	7.67E+06	639
NC	35	3.87E+05	9.70E+07	6773
ND	34	4.05E+03	8.72E+05	69
NE	39.14	1.15E+05	2.94E+07	2259
NH	77	4.20E+04	1.01E+07	1618
NJ	68	5.23E+05	1.38E+08	17767
NM	28	1.90E+05	3.99E+07	2663
NV	35	9.56E+03	2.53E+06	167
NY	44.69	2.15E+06	6.31E+08	47998
OH	32	2.08E+06	6.47E+08	33221

Continued

OK	18.5	1.63E+05	3.13E+07	1506
OR	35	5.91E+05	2.00E+08	10,339
PA	48.37	2.29E+06	6.68E+08	55,404
RI	52	5.04E+04	1.46E+07	1309
SC	35	2.59E+05	6.76E+07	4532
SD	39.5	6.71E+04	1.67E+07	1326
TN	34	7.50E+05	2.16E+08	12,748
TX	27.8	9.89E+05	2.69E+08	13,745
UT	37.82	1.67E+05	5.15E+07	3161
VA	40.35	6.89E+05	2.40E+08	13,903
VT	96	5.19E+03	1.21E+06	249
WA	52.65	4.85E+05	1.47E+08	12,758
WI	42.5	3.59E+05	9.77E+07	7635
WV	45.18	8.17E+04	2.27E+07	1847
WY	55	3.97E+04	1.11E+07	1091
Totals		2.78E+07	8.27E+09	559,645

4) The Input-Output Analysis Approach to Life Cycle Assessment

Because of the significant amount and diverse variety of products donated, measuring the embodied energy presented a formidable research challenge. After review of the lifecycle assessment literature including ISO 14040 and 14044 standards, we realized that the scope of this effort was unprecedented [65] [66] [67]. With the wide range of products and materials, conventional Life Cycle Assessment (LCA) processes were not deemed feasible. Therefore, a streamlined, four- step methodology was developed. This incorporates the estimation approach for products [68].

First, information on the embodied energy coefficients for primary raw materials and finished products was compiled, when available. The embodied energy coefficient was not available for most of the donated products. Since the products were composed of primary raw materials that went through a series of manufacturing processes, the bills of materials and the manufacturing processes for sampled products were collected from the product manufacturer's websites, and patent documents. Environmental Product Declarations and Climate Declarations, documents published by manufacturers from Europe and Japan for the users to make purchase decisions based on the environmental performance of the product, provided information on electric power, fossil fuel, and water; emissions into air and water; usage and maintenance phase energy and material consumption; and CO_2 emissions at production, usage and disposal phases [69]. The literature provided information on manufacturing processes including conventional metal manufacturing processes like machining, milling, grinding, casting, forging, and other processes in plastic, composite and nano-material manufacturing [66] [69]-[75].

Second, the information on the donated products from the Home Depot database was systematically sampled, reviewed, and then categorized, initially into 170 subcategories, and then further consolidated into 23 categories based on primary materials.

Third, detailed information for those products without coefficients from the first step was collected to fill in gaps in the data. This third step included evaluating the Bill of Materials and life cycle inventory resources based on Ganzheitliche Bilanz (GaBi) 4.0 databases to identify representative materials and process flows. The GaBi database (developed by PE International; http://www.gabi-software.com) is a modeling, reporting, and diagnostic software tool for LCA researchers and practitioners focusing on product sustainability performance during design and planning. The lifecycle inventory spans several industries and sectors.

Fourth, the embodied energy was calculated by assigning embodied energy coefficient values in MJ/kg to the appropriate SKUs in the given categories and multiplying them by the weight and quantity for each SKU. The calculated products were summed across categories in order to estimate the total embodied energy. The energy coefficients are commonly found in terms of MJ/kg (Mega Joules/kilogram), MJ/L, MJ/m^2, MJ/m^3, MMBtu/ft^2, and kWh/unit. All the units were converted to MJ/kg before doing the final analysis. To convert the non-standard units to MJ/kg, research on the average density of the products was necessary. This was found primarily from The Home Depot website (http://www.homedepot.com/) based on the products donated and calculated the average density. The final energy values are given in kWh (kilowatt-hours) and MT CO_2 Eq (Metric Tonnes CO_2 Equivalent). The U. S. EPA Interactive Units Convertor available under the Coalbed Methane Outreach Program was used to do unit conversions (https://www.epa.gov/cmop/units-converter).

*Embodied Energy = Weight * Total Donated Quantity * Embodied Energy Coefficient* (Equation (3))

Based on this process with the values summarized in **Table 3**, we estimated the total embodied energy to be **151,665,667 kWh (546 TJ)**.

6) The Streamlined Process Approach to Life Cycle Assessment

In an effort to further expand the embodied energy analysis, a second method was also used. The Economic Input-Output Life Cycle Assessment [76] approach estimates the materials, energy resources, and emissions associated with economic activities throughout the product supply chain. Originally theorized and developed by Nobel Laureate in Economic Sciences Wassily Leontief [77], the method was later advanced by researchers at the Green Design Institute of Carnegie Mellon University and made publicly available in 1995 via an online software tool [76]. It has been accessed widely and utilized for various products, services, and infrastructure systems [78]. Rather than focus on individual, ite

Table 3. Total Embodied Energy Values for Each Product Category (Estimation Method).

Category #	Category Name	Total Weight (tons)	Total Embodied Energy (kWh)
1	Additives & Adhesions	4.49E+01	6.78E+05
2	Batteries	1.82E+00	3.40E+04
3	Cement Based Products	2.99E+02	5.69E+05
4	Ceramic Materials	1.77E+03	2.89E+06
5	Construction Chemicals	6.52E+01	2.88E+05
6	Construction Materials	5.55E+02	1.99E+06
7	Construction Natural Materials	3.37E+02	5.95E+05
8	Electrical & Electronic Products	1.19E+03	1.68E+07
9	Electrical Cables & Wires	6.02E+01	3.95E+05
10	Fiberglass	1.41E+02	1.20E+06
11	Glass Products	1.39E+03	7.63E+06
12	Metal Plastic Products	3.97E+02	4.18E+06
13	Metal Plastic Wood Products	1.66E+02	8.17E+06
14	Metal Products	1.78E+03	4.18E+06
15	Metal Rubber Products	8.10E+00	2.17E+05
16	Natural Products	1.85E+03	3.12E+07
17	Organic Products	3.31E+02	3.22E+05
18	Paints	8.20E+00	1.76E+05
19	Paper Cardboard Products	3.97E+00	2.43E+04
20	Plastic Products	1.11E+03	2.52E+07
21	Rubber Products	4.17E+01	1.12E+07
22	Wood Furniture	8.96E+02	2.23E+06
23	Wood Products	1.47E+03	1.93E+07

mize, and detailed production processes for each step of product development like process-based LCA approaches, which can become convoluted and difficult for complex systems, the EIO-LCA addresses comprehensive economies involving aggregated industry sectors. This is framed by broad economic boundary conditions which lack the specificity of process-based models [76]. The EIO-LCA models available for selection within the tool apply the EIO-LCA method to various national and state economies. Each model is comprised of national economic input-output models coupled with publicly available resource use and emissions data. Most nations create economic input-output models of their given economies to varying degrees of specificity and frequency. The U.S. EIO models (benchmark references) are created every five years and represent the transactions among some 400 industry sectors using the North American Industry Classification System (NAICS) or other generic categories. Uncertainties with data sources that are outdated or incomplete, and due to aggregated levels have been noted [76]. Findings from using the EIO-LCA tool provide insight into the relative impacts of different types of products, materials, services, or industries from resource extraction through final assembly. For this calculation we used the previously designated 170 product subcategories which represented the level of detail necessary to recode the products based on the EIO-LCA sectors. Based on these, 67 sectors were identified and included in the calculation in this EIO-LCA sectorial analysis. Using the EIO-LCA codes, and including the total number of products in each identified sector and the associated monetary values ($64.7 million total) retrieved from the original Framing Hope database, values were input into the EIO-LCA model. Steps in the EIO-LCA process include choosing the model of the options provided, selecting the industry and the sector, selecting the level of economic activity, and selecting the effects to display, after which the model can be run. As an example of sectorial results, for a single run the model chosen was the US National Producer Price Model for 2002, the industry and sector chosen was "other plastics production manufacturing" and the value of the sectorial economic activity was put in as $9,409,810, based on the summed values of the products donated in that category. For these model choices, the total estimated embodied energy for the "other plastics production manufacturing" sector was 140 TJ. **Figure 5** provides an example of the results for this largest embodied energy category.

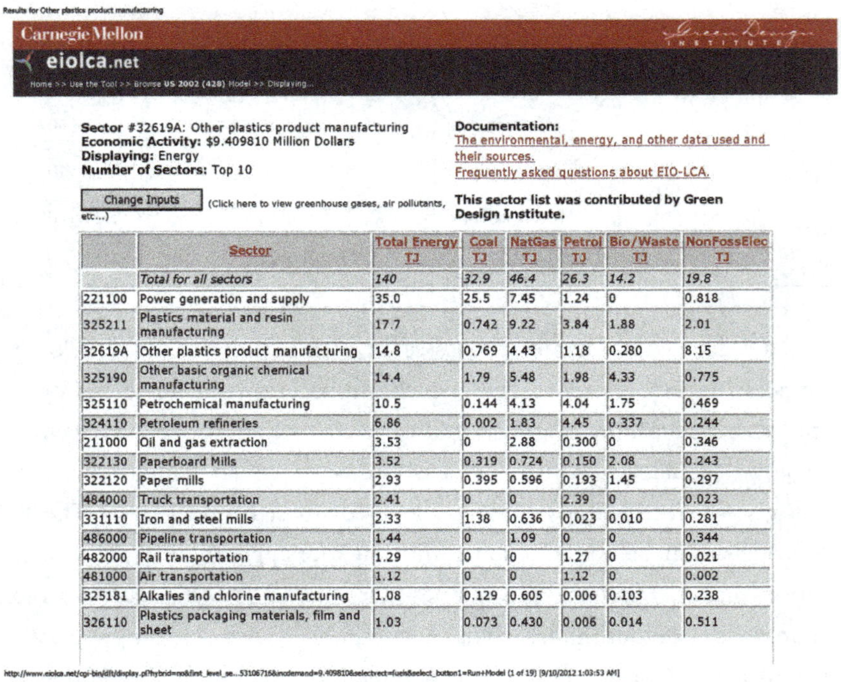

Results for Other plastics product manufacturing

Carnegie Mellon

eiolca.net

Green Design Institute

Home >> Use the Tool >> Browse **US 2002 (428)** Model >> Displaying...

Sector #32619A: Other plastics product manufacturing
Economic Activity: $9.409810 Million Dollars
Displaying: Energy
Number of Sectors: Top 10

Change Inputs (Click here to view greenhouse gases, air pollutants, etc...)

Documentation:
The environmental, energy, and other data used and their sources.
Frequently asked questions about EIO-LCA.

This sector list was contributed by Green Design Institute.

	Sector	Total Energy TJ	Coal TJ	NatGas TJ	Petrol TJ	Bio/Waste TJ	NonFossElec TJ
	Total for all sectors	*140*	*32.9*	*46.4*	*26.3*	*14.2*	*19.8*
221100	Power generation and supply	35.0	25.5	7.45	1.24	0	0.818
325211	Plastics material and resin manufacturing	17.7	0.742	9.22	3.84	1.88	2.01
32619A	Other plastics product manufacturing	14.8	0.769	4.43	1.18	0.280	8.15
325190	Other basic organic chemical manufacturing	14.4	1.79	5.48	1.98	4.33	0.775
325110	Petrochemical manufacturing	10.5	0.144	4.13	4.04	1.75	0.469
324110	Petroleum refineries	6.86	0.002	1.83	4.45	0.337	0.244
211000	Oil and gas extraction	3.53	0	2.88	0.300	0	0.346
322130	Paperboard Mills	3.52	0.319	0.724	0.150	2.08	0.243
322120	Paper mills	2.93	0.395	0.596	0.193	1.45	0.297
484000	Truck transportation	2.41	0	0	2.39	0	0.023
331110	Iron and steel mills	2.33	1.38	0.636	0.023	0.010	0.281
486000	Pipeline transportation	1.44	0	1.09	0	0	0.344
482000	Rail transportation	1.29	0	0	1.27	0	0.021
481000	Air transportation	1.12	0	0	1.12	0	0.002
325181	Alkalies and chlorine manufacturing	1.08	0.129	0.605	0.006	0.103	0.238
326110	Plastics packaging materials, film and sheet	1.03	0.073	0.430	0.006	0.014	0.511

http://www.eiolca.net/cgi-bin/dft/display.pl?hybrid=no&first_level_se...53106716&incdemand=9.409810&selectvect=fuels&select_button1=Run+Model (1 of 19) [9/10/2012 1:03:53 AM]

Figure 5. Embodied Energy EIO-LCA Example for "Other plastics product manufacturing" category.

The model was run for each of the 67 selected EIO-LCA codes encompassing the range of donated products. A list of contributing industrial sectors and energy sources was output by the model. The total embedded energy for each run was recorded and summed for all of the product codes. Based on this approach, the total embodied energy was calculated to be **259,670,000 kWh (935 TJ).**

7) Question #3: What is the Energy Conserved through Use of the Donated Products?

Most of the end users in the Framing Hope program qualify as low income and are in need of relief from the burden of high-energy costs. In fact, energy costs can pose serious hurdles to financial security for low-income families. Stakeholders in the lowest quintile spend about twice as much on energy as people in the highest quintile [77]. This energy cost disparity between socioeconomic groups is likely to increase as energy costs rise. In terms of energy efficient products, an important development in the United States has been implementation of the Energy Star program in 1992. This voluntary labeling program has been highly trusted and successful with over 20,000 private and public sector organizations. Studies of estimated savings based on the program's products vis-à-vis conventional products have found substantial energy and economic benefits [78].

As noted above, information from the Home Depot database allowed us to review and compile specific details for the donated products. With respect to establishing the energy conservation potential and energy consumption, some of the products had energy values associated with usage but most did not. By comparing the energy efficiencies of select donated products to those that were typically used by recipients of nonprofit services, estimates were determined for the energy savings attributable to the use of products. For calculating the energy savings of the Framing Hope Program, the *Framing Hope Partner Survey* [34] was used to make assumptions about the percentage of new products that would have otherwise been bought should the program have not existed. The survey provided feedback from participating nonprofit organizations regarding the usefulness of the products.

The total number of products initially examined for energy savings was 265,516 out of the nearly 4.7 million products donated. Estimates on energy usage for individual products were derived for 19,268 donated products, most of which were Energy Star certified. This allowed use of Energy Star program standards, data, and calculator resources, as developed by USDOE and USEPA (http://www.energystar.gov/). The main product categories investigated included appliances, ceiling fans, doors, insulation, light bulbs, light fixtures, skylights, and windows. Each category was individually analyzed. Estimates were calculated for behavioral changes related to a subset of donated products which could be directly linked to energy saving behaviors, like using less air conditioning when a ceiling fan is installed. These energy savings accrue when the usage or installation of the products result in a reduction in energy expenditures for the consumer. Specific products that fall into this category include insulation, windows, doors, and ceiling fans. Important considerations in the behavioral analysis included whether the products were used in retrofitting or new construction projects as well as geographical region.

A summary of products, energy savings, and monetary values are included in **Table 4.** Based on the analysis of the individual categories of products and range of values, the annual consumption of energy savings was estimated to be **3,378,319 kWh** (estimated value of **$645,761**). Over the lifetime of the products, the energy savings were estimated to be **41,067,206 kWh** (estimated value of **$5,408,924**).

5. Discussion and Conclusions

This effort provides the first analysis of the potential sustainability-related benefits that accrue to a corporate product philanthropy program. The benefits include corporate as well as community and global benefits. The main corporate benefits are the cost savings that come from reduced landfill tipping fees, but also include a positive community image as well as tax savings. The community benefits are more varied, and include saved landfill space and reduced energy costs due to use of more energy efficient appliances and products as well as the reduced energy costs that result from using products, like insulation and ceiling fans, that encourage more energy efficient home heating and cooling behaviors. The global benefits result from the energy saved by reducing the amount of products manufactured when products are used rather than thrown in a landfill

Table 4. Summary of Product Categories, Total Annual Energy Savings, and Monetary Savings.

Product Category	Annual Energy Savings (kWh)	Annual Monetary Savings ($)
Clothes Washer (Electric/Gas)	5330/2345	613/270
Dishwashers (Electric/Gas)	11985/2023	1378/233
Dehumidifiers	10230	1176
Air Purifiers	12436	1430
Ceiling Fans (Technical/Behavioral)	140078/6570536	17451/1028426
Doors	39808	4776
Insulation	78292	9000
Light Bulbs	238260	26321
Light Fixtures	588370	64897
Skylights	6058	697
Windows	38524/168786	4686/22140

*For ceiling fans, technical savings are based on energy efficiencies of Energy Star versus non-Energy Star products; and behavioral savings include savings from setting air conditioning thermostat higher (~4 degree difference).

when the products reach the end of their salable or marketable "shelf life."

These results provide an indication of the relative magnitude of environmental, economic, and energy benefits associated with the array of donated products in the Framing Hope program. With the shift in corporate policy and operations, more than 30,000 yd^3 of landfill space was saved and over $700,000 of waste tipping fees were not spent. Calculations based on both national averages and more refined state level data demonstrate that significant savings were derived from the Framing Hope program. Considering the concerns associated with landfilling of solid waste including hazards associated with fire, potential disease vectors (animals and insects), uncontrolled release of gases, emissions of greenhouse gases, releases of leachate that have the potential for adverse effects on surface and groundwater resources, increased levels of electronic waste, odor, and aesthetics [54] [79] [80], programs such as Framing Hope represent important interventions toward increased public safety, improved environmental health, and more sustainable communities.

Two lifecycle approaches were used to calculate the embodied energy associated with the donated products. The modified LCA process involved far more assumptions and uncertainties with respect to underlying materials and production processes compared to the economic input output technique whereby the product data coding was closely aligned with the sector categories contained in the EIO-LCA model. Although the results are varied by somewhat less than a factor of two between the two methods, these differences are not unexpected based on the different levels of data confidence. There is value added in measuring and analyzing energy from disparate perspectives. Both approaches determined that significant energy was embedded in the donated products representing a substantial opportunity cost linked to product management decisions.

Of the three research questions, evaluations of avoided energy effects presented the greatest challenge. This was partly due to the relatively small number of products that were certified as energy efficient (*i.e.*, Energy Star) with well-defined calculators and guidance to determine energy savings. Additionally, metric uncertainties with respect to individual level behaviors and use, types of construction projects, and geographic regions presented difficult complications. Thus, the calculations and results obtained were a positive step but further investigation is suggested.

These findings demonstrate that the Framing Hope program has had a positive impact on community sustainability. With the enhanced focus on lifecycle product management and resource efficiency along the supply chain for retail business, partnerships with local nonprofits are able benefits of corporate programs.

Acknowledgements

This research has been partially funded by Indiana University Office of Sustainability (IUOS), Environmental Research & Education Foundation (EREF), Kansas State University Center for Hazardous Substance Research (CHSR), and Consortium for Environmental Stewardship and Sustainability (CESAS).

References

[1] Heuer, M. (2011) Ecosystem Cross-Sector Collaboration: Conceptualizing an Adaptive Approach to Sustainability Governance. *Business Strategy and the Environment*, **20**, 211-221. https://doi.org/10.1002/bse.673

[2] Hubbard, G. (2009) Measuring Organizational Performance: Beyond the Triple Bottom Line. *Business Strategy and the Environment*, **18**, 177-191. https://doi.org/10.1002/bse.564

[3] Portney, K. and Cuttler, Z. (2010) The Local Nonprofit Sector and the Pursuit of Sustainability in American Cities: A Preliminary Exploration. *Local Environment*, **15**, 323-339. https://doi.org/10.1080/13549831003677704

[4] Sachs, J.D. (2015), The Age of Sustainable Development. Columbia University Press, New York. https://doi.org/10.7312/sach17314

[5] Waddell, S. (2000) A "Win-Win" Role for Civil Society in Business Strategy. *Journal for Nonprofit Management*, **4**, 24-43.

[6] Loza, J. (2004) Business-Community Partnerships: The Case for Community Organization Capacity Building. *The Journal of Business Ethics*, **53**, 297-311. https://doi.org/10.1023/B:BUSI.0000039415.90007.56

[7] Austin, J. (2000) Strategic Collaboration between Nonprofits and Businesses. *Nonprofit and Voluntary Sector Quarterly*, **29**, 69-97. https://doi.org/10.1177/0899764000291S004

[8] Chaskin, R., Brown, P., Venkatesh, S. and Vidal, A. (2001) Building Community Capacity. Aldine De Gruyter, New York.

[9] Delfin, F.G. and Tang, S. (2008) Foundation Impact on Environmental Nongovernmental Organizations: The Grantees' Perspective. *Nonprofit and Voluntary Sector Quarterly*, **37**, 603-625. https://doi.org/10.1177/0899764007312667

[10] Kelly, C.C.M (2011) Understanding Accountability in Social Enterprise Organisations: A Framework. *Social Enterprise Journal*, **7**, 224-237. https://doi.org/10.1108/17508611111182386

[11] Milen, A. (2001) What Do We Know about Capacity Building: An Overview of Existing Knowledge and Good Practice. Department of Health Service Pro Vision. World Health Organization, Geneva.

[12] Nikolaou, I.E., Evangelinos, K.I. and Allan, A. (2013) A Reverse Logistics Social Responsibility Evaluation Framework Based on the Triple Bottom Line Approach. *Journal of Cleaner Production*, **56**, 173-184.

[13] Perrini, F., Russo, A., Tencati, A. and Vurro, C. (2011) Deconstructing the Relationship between Corporate Social and Financial Performance. *Journal of Business Ethics*, **102**, 59-76. https://doi.org/10.1007/s10551-011-1194-1

[14] Phillips, W., Lee, H., Ghobadian, A., O'Regan, N. and James, P. (2014) Social Innovation and Social Entrepreneurship A Systematic Review. *Group and Organization Management*, **40**, 428-461.

[15] Zhang, D.D. and Swanson, L.A. (2013) Social Entrepreneurship in Nonprofit Organizations: An Empirical Investigation of the Synergy between Social and Business Objectives. *Journal of Nonprofit and Public Sector Marketing*, **25**, 105-125.

https://doi.org/10.1080/10495142.2013.759822

[16] Agyeman, J. and Evans, B. (2003) Toward Just Sustainability in Urban Communities: Building Equity Rights with Sustainable Solutions. *The Annals of the American Academy of Political and Social Science*, **590**, 35-53. https://doi.org/10.1177/0002716203256565

[17] Bashour, H., Bellows, L., Krishnan, B., LaBeau, K., Lindeen, L., Rog, V., Smothers, L., Summers, L., Wendelin, M.B. and Vahnie, E. (2010) Assessing Product Philanthropy's Impact on Community Sustainability: Energy and Solid Waste. Indiana University School of Public and Environmental Affairs Capstone Papers.

[18] Carnegie, A. (1899) Wealth. *The North American Review*, **148**, 653-664.

[19] Friedman, M. (1970) The Social Responsibility of Business Is to Increase Its Profits. New York Times Magazine, September 13, 32-33.

[20] Young, D.R. and Burlingame, D.F. (1996) Paradigm Lost: Research toward a New Understanding of Corporate Philanthropy. Corporate Philanthropy at the Crossroads, 158-176.

[21] Illingworth, P., Pogge, T. and Wenar, L. (2011) Giving Well: The Ethics of Philanthropy. Oxford University Press, New York. https://doi.org/10.1093/acprof:oso/9780199739073.001.0001

[22] Johnson, O. (1966) Corporate Philanthropy: An Analysis of Corporate Contributions. *The Journal of Business*, **39**, 489-504. https://doi.org/10.1086/294890

[23] Wang, H. and Qian, C. (2011) Corporate Philanthropy and Corporate Financial Performance: The Roles of Stakeholder Response and Political Access. *Academy of Management Journal*, **54**, 1159-1181. https://doi.org/10.5465/amj.2009.0548

[24] Bowen, H.R. (1953) Social Responsibilities of the Businessman. Harper and Row, New York.

[25] Carroll, A. (1991) The Pyramid of Corporate Social Responsibility: Toward the Moral Management of Organizational Stakeholders. *Business Horizons*, **34**, 39-48.

[26] Brammer, S. and Millington, A. (2005) Corporate Reputation and Philanthropy: An Empirical Analysis. *Journal of Business Ethics*, **61**, 29-44. https://doi.org/10.1007/s10551-005-7443-4

[27] Chen, J.C., Patten, D. and Roberts, R. (2008) Corporate Charitable Contributions: A Corporate Social Performance or Legitimacy Strategy? *Journal of Business Ethics*, **82**, 131-144.

Foster, M.K., Meinhard, A.G., Berger, I.E. and Krpan, P. (2009) Corporate Philanthropy in the Canadian Context: From Damage Control to Improving Society. *Nonprofit and Voluntary Sector Quarterly*, **38**, 441-466.

[28] Halme, M. and Laurila, J. (2009) Philanthropy, Integration or Innovation? Exploring the Financial and Societal Outcomes of Different Types of Corporate Responsibility. *Journal of Business Ethics*, **84**, 325-339. https://doi.org/10.1007/s10551-008-9712-5

[29] Nicholls, A. (2009) "We Do Good Things, Don't We?": "Blended Value Accounting" in Social Entrepreneurship. *Accounting, Organizations and Society*, **34**, 755-769.

[30] Patten, D. (2008) Does the Market Value Corporate Philanthropy? Evidence from the Response to the 2004 Tsunami Relief Effort. *Journal of Business Ethics*, **81**, 599-607. https://doi.org/10.1007/s10551-007-9534-x

[31] Ricks, J. and Williams, J. (2005) Strategic Corporate Philanthropy: Addressing Frontline Talent Needs through an Educational Giving Program. *Journal of Busi-*

ness Ethics, **60**, 147-157. https://doi.org/10.1007/s10551-005-1175-3

[32] Bhattacharya, C., Sen, S. and Korschun, D. (2008) Using Corporate Social Responsibility to Win the War for Talent. *MIT Sloan Management Review*, **49**, 37-44.

[33] Gazley, B. and Abner, G. (2014) Evaluating a Product Donation Program: Challenges for Charitable Capacity. *Nonprofit Management and Leadership*, **24**, 337-355. https://doi.org/10.1002/nml.21094

[34] Luo, X. and Bhattacharya, C. (2006) Corporate Social Responsibility, Customer Satisfaction, and Market Value. *Journal of Marketing*, **70**, 1-18. https://doi.org/10.1509/jmkg.70.4.1

[35] Nidumolu, R., Prahlad, C. and Rangaswami, M. (2009) Why Sustainability Is Now the Key Driver of Innovation. *Harvard Business Review*, **87**, 56-64.

[36] Porter, M. and Kramer, M. (2011) Creating Shared Value. *Harvard Business Review*, **89**, 62-77.

[37] Austin, J. and Seitanidi, M. (2012) Collaborative Value Creation: A Review of Partnering between Nonprofits and Businesses: Part I. Value Creation Spectrum and Collaboration Stages. *Nonprofit and Voluntary Sector Quarterly*, **41**, 726-758. https://doi.org/10.1177/0899764012450777

[38] Harris, M. (2012) Nonprofits and Business: Toward a Subfield of Nonprofit Studies. *Nonprofit and Voluntary Sector Quarterly*, **41**, 892-902. https://doi.org/10.1177/0899764012443735

[39] National Center for Charitable Statistics (NCSS) (2010) The Nonprofit Almanac 2010: Number of Nonprofit Organizations by State. Washington DC Urban Institute.

[40] Giving USA (2012) The Annual Report on Philanthropy for the Year 2011. The Center on Philanthropy at Indiana University.

[41] Jackson, P.J. (2006) Charitable Contributions of Food Inventory: Proposals for change. Congressional Research Service: Library of Congress.

[42] Chronicle of Philanthropy (2012) Special Report: How American Gives. The Chronicle of Philanthropy, 24, August 23, 2012. https://www.philanthropy.com/specialreport/how-america-gives-2014/1?CID=cpw_directory_data

[43] Lim, T. (2009) Measuring the Value of Corporate Philanthropy: Social Impact, Business Benefits, and Investor Returns. Committee for Encouraging Corporate Philanthropy, New York.

[44] Committee Encouraging Corporate Philanthropy (CECP) (2011) Giving in Numbers 2010 Edition. Corporate Giving Standard, New York.

[45] Ross, J. and McGiverin-Bohan, K. (2011) The Business Case for Product Philanthropy. Indiana University School of Public and Environmental Affairs.

[46] Good360 (2012) The Framing Hope Product Donation Program: Partnering Home Depot Stores with Local Non Profit Organizations. The Home Depot Foundation, Atlanta.

[47] Hansen, S., Dunford, B., Boss, A., Boss, R. and Angermeier, I. (2011) Corporate Social Responsibility and the Benefits of Employee Trust: A Cross-Disciplinary Perspective. *Journal of Business Ethics*, **102**, 29-45. https://doi.org/10.1007/s10551-011-0903-0

[48] Edelman (2010) Citizens Engage! Edelman Goodpurpose Study 2010 4th Annual Global Consumer Survey.

[49] Williams, R. and Barrett, J. (2000) Corporate Philanthropy, Criminal Activity, and Firm Reputation: Is there a link? *Journal of Business Ethics*, **26**, 341-350. https://doi.org/10.1023/A:1006282312238

[50] Paarlberg, L.E. and Meinhold, S. (2012) Using Institutional Theory to Explore Local Variations in United Way's Community Impact Model. *Nonprofit and Voluntary Sector Quarterly*, **41**, 826-849. https://doi.org/10.1177/0899764011418123

[51] U.S. Census Bureau (2009) Consumer Expenditure Survey 2009. Collected by the U.S. Bureau of Labor Statistics. http://www.bls.gov/cex/

[52] Seefeldt, K. and Abner, G. (2011) Framing Hope End User Pilot Study. Indiana University School of Public and Environmental Affairs. Indiana University.

[53] U.S. Environmental Protection Agency (USEPA) (2011) Municipal Solid Waste Generation, Recycling, and Disposal in the United States: Facts and figures for 2010. Solid Waste and Emergency Response. EPA-530-F-11-005.

[54] Clark, C., Jambeck, J. and Townsend, T. (2006) A Review of Construction and Demolition Debris Regulations in the United States. *Critical Reviews in Environmental Science and Technology*, **36**, 141-186. https://doi.org/10.1080/10643380500531197

[55] U.S. Environmental Protection Agency (USEPA) (1998) Characterization of Building-Related Construction and Demolition Debris in the United States. Office of Solid Waste. EPA530-R-98-010.

[56] Agency for Toxic Substances and Disease Registry (ATSDR) (2003) Exposure Investigation: Hydrogen Sulfide Exposure in Warren Township, Trumbull County, Ohio. U.S. Department of Health and Human Services, Public Health Service, Atlanta.

[57] Agency for Toxic Substances and Disease Registry (ATSDR) (2014) Toxicological Profile for Hydrogen Sulfide/Carbonyl Sulfide (Draft for Public Comment). U.S. Department of Health and Human Services, Public Health Service, Atlanta.

[58] Colledge, M. and Wilder, L. (2008) Construction and Demolition (CandD) Landfills: Emerging Public and Occupational Health Issues. *Journal of Environmental Health*, **71**, 50-52.

[59] Dixon, N. and Jones, D.R.V. (2005) Engineering Properties of Municipal Solid Waste. *Geotextiles and Geomembranes*, **23**, 205-233.

[60] Hudson, A.P., White, J.K., Beaven, R.P. and Powrie, W. (2004) Modelling the Compression Behaviour of Landfilled Domestic Waste. *Waste Management*, **24**, 259-269.

[61] Machado, S.L., Carvalho, M.F. and Vilar, O.M. (2002) Constitutive Model for Municipal Solid Waste. *Journal of Geotechnical and Geoenvironmental Engineering*, **128**, 940-951. https://doi.org/10.1061/(ASCE)1090-0241(2002)128:11(940)

[62] Van Haaren, R., Goldstein, N. and Themelis, N.J. (2010) The State of Garbage in America. *BioCycle*, **49**, 16-23.

[63] Waste360.com (2010).

[64] Costanza, R. (1980) Embodied Energy and Economic Values. *Science*, **210**, 1219-1224. https://doi.org/10.1126/science.210.4475.1219

[65] Sahni, S., Boustani, A., Gutowsk, T. and Graves, S. (2010) Furniture Remanufacturing and Energy Savings. Environmentally Benign Laboratory. Laboratory for Manufacturing and Productivity. Sloan School of Management. MITEI-1-e-2010.

[66] U.S. Environmental Protection Agency (USEPA) (2008) Wire and Cable Insulation and Jacketing: Life-Cycle Assessments for Selected Application. Office of Pollution Prevention and Toxics, Design for the Environment. EPA-744-S-08-001.

[67] Ricoh Co. Ltd. (2000) Environmental Product Declaration Ricoh Imagio MF6550

Digital Copier Main Unit. Registration Number: S-EP-00004.
http://www.environdec.com/en/Articles/EPD/Ricoh-first-with-a-global-Type-III-environmental-declaration/

[68] Duque Ciceri, N., Gutowski, T.G. and Garetti, M. (2010) A Tool to Estimate Materials and Manufacturing Energy for a Product. *IEEE International Symposium on Sustainable Systems and Technology*, 1-6.
http://ieeexplore.ieee.org/xpl/mostRecentIssue.jsp?punumber=5496818

[69] Blanchard, S. and Reppe, P. (1998) Life Cycle Analysis of a Residential Home in Michigan. Center for Sustainable System and School of Natural Resources and Environment, University of Michigan. http://www.p2pays.org/ref/37/36507.pdf

[70] Grondzik, W., Kwok, A., Stein, B. and Reynolds, J. (2010) Mechanical and Electrical Equipment for Buildings. John Wiley and Sons, Hoboken.

[71] Lawson, B. and Rudder, D. (1996) Building Materials Energy and the Environment: Towards Ecologically Sustainable Development. Royal Australian Institute of Architects.

[72] Nishimura, K., Hondo, H. and Uchiyama, Y. (1995) Derivation of Energy-Embodiment Functions to Estimate the Embodied Energy from the Material Content. *Energy*, **21**, 1247-1250.

[73] Procurement, G.P. (2010) Thermal Insulation Technical Background Report. European Commission.
http://www.vergabebrief.de/wp-content/uploads/2014/09/Background_Report_Thermal-Insulation.pdf

[74] U.S. Department of Energy (USDOE) (2009) Buildings Energy Data Book (2009 BEDB). Office of Energy Efficiency and Renewable Energy.
http://buildingsdatabook.eren.doe.gov/DataBooks.aspx

[75] Carnegie Mellon University Green Design Institute (CMU-GDI) (2008) Economic Input-Output Life Cycle Assessment (EIO-LCA), US 1997 Industry Benchmark Model. http://www.eiolca.net

[76] Leontief, W. (1938) The Significance of Marxian Economics for Present-Day Economic Theory. *The American Economic Review*, **28**, 1-9.

[77] Hendrickson, C., Lave, L. and Matthews, H. (2006) Environmental Life Cycle Assessment of Goods and Services: An Input-Output Approach. Resources for the Future Press, Washington DC.

Podesta, J. and Miller, S. (2009) Seizing the Energy Opportunity. Pathways, Fall, 14-18.

[78] Sanchez, M., Webber, C., Brown, R. and Homan, G. (2008) Status Report. Savings Estimates for the ENERGY STAR® Voluntary Labeling Program. Lawrence Berkeley National Laboratory. LBNL-56380, Berkeley.

[79] Tchobanoglous, G., Thiesen, H. and Vigil, S. (1993) Integrated Solid Waste Management: Engineering Principles and Management Issues. McGraw-Hill, New York.

[80] American Society of Civil Engineers (ACSE) (2009) Report Card for America's Infrastructure, Water and Environment—Solid Waste. ASCE Press, Reston.

11

Residual Effects from Occupational Mercury Exposure Include a Proposed Mercury Tremor Biomarker or "Fingerprint"

Linda Jones

School of Psychology, Massey University, Wellington, New Zealand
Email: L.M.Jones@massey.ac.nz

Abstract

The study investigated residual effects of high levels of occupational mercury exposure, 30 years after a cohort of women worked in public service dentistry. They had all used copper amalgam in a pellet form that required heating and handling, and silver amalgam before the encapsulated form was available. Mercury handling practices changed in the mid-1970 when the workforce was urine tested and mercury poisoning became apparent. The aim was to compare control group and exposed group scores on tasks from a neurobehavioural test battery; plus survey results from a composite health, work history and environmental influences survey. The findings showed that the exposed and control groups were equivalent not only on those variables that one would want to be matched (age, alcohol consumption), but also on many of the cognitive and psychomotor test scores. The present paper focuses on psychomotor skill and tremor patterns. Tremor patterns were seen as generating new evidence of long term effects of the historic mercury insult. Data also suggest that there may be a distinctive mercury "fingerprint", in samples of sinusoidal waveforms that may have potential as a non-invasive sub-clinical biomarker for adverse effects of mercury exposure, in screening or workplace monitoring.

Keywords

Mercury, Tremor, Dentistry, Women, Psychomotor Tests

1. Introduction

In New Zealand in the 1960s and 1970s every primary school had a state funded dental clinic, staffed by women who were specifically trained in a limited range of conservative dentistry techniques for treating children. The filling materials were copper amalgam for deciduous teeth, and silver amalgam for permanent teeth. By the mid-1970s it was found that 10% of the staff had urine-mercury levels over 50 ug/l [1] [2]. Many had been medically diagnosed with mood disorders that were incorrectly attributed the notion that women were emotionally labile, rather than suffering from mercury poisoning[1]. Copper amalgam was withdrawn from use in 1975 and although the use of silver amalgam has persisted, an encapsulated form was introduced with mechanical mixing. This has created a natural experimental design for investigating three little-reported aspects of occupational mercury safety: possible long term effects, effects on woman, and studies of metals including the heavy metal, mercury, in dental materials generally.

One historical difficulty for a retrospective study is that although a total working population sample of 1300 women were urine tested in one year, few were given their results. There was no treatment for women with elevated results, only enforced time off work. The government agency responsible for urine-testing was not required to produce a report, and only minimal information on the episode can be found in Department of Health archived records. Hence it is not possible to produce a model that quantifies individual mercury exposure and residual effects. However a model of exposure can be constructed from four other variables: a description of mercury use in the New Zealand School Dental Service at the time; the volume of amalgam use from filling placement data in Department of Health's annual reports; a description of the clinic environment and gazetted instructions to the women about mercury hygiene; and finally, survey responses on the women's recall of working with mercury.

Description of Mercury use. Copper amalgam, 70% mercury to 30% copper, was supplied in pellet form. Pellets required heating over a flame to soften the copper and release the mercury so the amalgam could be mixed (see **Figure 1**).

Figure 1. Heated copper amalgam pellets with mercury released, ready for trituration using a glass mortar and pestle.

[1]Newspaper quote from Dr B Glass, occupational physician who tested the women.

The mercury vapor levels in the ambient air peaked at these times. Frykholm [3] reported on experimental testing of air levels in such a situation, and suggested it was possible to reach 3000 μg Hg/m^3 in air around the burner. He suggested that copper amalgam could only safely be used with a ventilation and exhaust system. In the New Zealand scenario, the burner was close to both the operator and patient. There was no assisted ventilation. After mixing to reincorporate the copper and mercury, the amalgam was scooped up by hand using a cotton "squeeze cloth" and excess mercury was wrung out over the mortar. Droplet spills were common.

Silver amalgam was supplied in the separate ingredients of silver alloy powder and liquid mercury. These were weighed on a small balance scale in the ratio of five parts silver six parts mercury, with the mercury poured from a wooden "dropper" bottle. Frykholm's experiments described this situation also, reporting 400 μg Hg/m^3 in the air above the silver amalgam mix [3]. As with the copper mix, the powdered silver and room temperature mercury were mixed to a paste in the mortar and pestle, and again when the desired consistency was reached, excess mercury was wrung out by hand. Waste amalgam and mercury from the mortar were placed in water filled and lidded glass jars for future collection and recycling.

Peak mercury in ambient air level as measured by Frykholm [3] shows there was a constant and continual risk of mercury poisoning for the women, when compared with the generally accepted international value is 25 μg/m^3 [4] or the German MAK[2] value of 10 μg/m^3 [5].

Volume of amalgam use. There were just over 1300 women, mostly in full time positions, staffing school dental clinics in the years 1974 to 1976. That each prepared on average 10 amalgam fillings per day can be extrapolated from data in the official Department of Health annual report to government (**Table 1**). This showed that collectively the women performed more than 2 million fillings during the 44 weeks per year when clinics were open. This would require hand contact with mercury on at least 10 occasions per day, and a constant high level of ambient air mercury.

Dental clinic environments. School dental clinics were built to a standard design with one large window to let in light, but it did not open. Smaller side-wall windows could be. Variation and fluctuations in New Zealand's seasonal

Table 1. Filling Output, by Staff and Year.

Year	Number of Staff	Total Number of Fillings	Mean per Person per Year
1974	1.356	2.589.019	1.909
1975	1.371	2.248.326	1.640
1976	1.319	2.202.169	1.670

Source: AJHR, 1975, 1976.

[2]Maximale Arbeitsplatz-Konzentration or maximum concentration in air.

temperatures meant that the departmental instruction for routine mercury risk prevention—to keep windows open—was impossible to follow [6]. Women reported that in cold weather the clinic windows were kept closed and heaters turned on. There was no air conditioning or exhaust ventilation system. Women were not trained to wear, nor supplied with, protective clothing such as gloves or masks; and were sufficiently uninformed about mercury toxicity that they gave spilled droplets of mercury to their child patients in empty burr boxes as a reward for cooperative behavior [7].

The high levels and frequent peaks of mercury in air, the manual measurement of liquid mercury for silver fillings, the naked-hand expression of excess mercury from amalgam mixes and clean-up of accidental spills, the absence of any protective clothing, and the sheer volume of work, demonstrates that even the most conscientious woman would be at risk of mercury poisoning in this occupational environment.

The neurobehavioural effects of mercury exposure are well documented for people who have chronic or continued exposure. These include psychosomatic illness experience; and deficits or changes in affect, cognitive skills, psychomotor skills and sensory function. Meta-analysis has established exposure and test performance relationships for different domains with common neurobehavioural tests [5]. Studies specifically in dentistry have found adverse neurobehavioral effects from chronic low-level mercury exposure [8] [9] [10] [11] [12]. From these there is no clear evidence from which to predict the long-term health outcomes for dentists who may at one time have experienced neurobehavioural changes from occupational mercury exposure.

There are few longitudinal studies of mercury exposed workers, and in those studies findings are not consistent. [13] [14] [15] [16]. Ratcliffe *et al.* critically assessed the methodology and analysis of 164 mercury exposure studies and concluded that "the evidence for association between neurologic effects and inorganic mercury is irrefutable" p. 238, [17], yet some uncertainty remains about long term effects once exposure ceases. Enduring health and psychomotor function changes are seen in some chloralkali factory workers, but the possibility remains that other deficits are reversible once exposure ceases [18].

In addition, women have been neglected in neurobehavioural studies of occupational mercury exposure. There are three reports specifically of women and occupational mercury exposure and a limited number of studies that include gender more broadly. In total, at the time data were collected for the present study, there were neurobehavioural data on fewer than 30 women [11] [19] [20].

The aim of the present paper is to report on psychomotor assessment, including tremor analysis, from a neurobehavioural test battery, with women who were occupationally exposed to mercury as young women, at least 30 years prior to their assessment, and matched controls with no history of occupational exposure to heavy metals (Findings from other tests are reported elsewhere [21]). Further, tremor analysis is explored as a potential non-invasive, early warning technique for people currently occupationally exposed to mercury in any form.

2. Materials and Method

Participants were approached by mail with a request to volunteer with a matched control. The mercury exposed group was 43 women ($\bar{x}$ 2.2 years, SD 1.2). Inclusion criteria included graduation between 1968 and 1971 with a minimum of three year's work experience with the New Zealand school dental service. The response rate was 40.2%. The control group were 13 sisters and 19 women friends (total 32) of the mercury-exposed group ($\bar{x}$ 51.4 years, SD 4.5). The power analysis formula was that recommended in the *Adult Environmental Neurobehavioral Test Battery* [22] to detect a 20% between groups difference across the psychomotor tests.

Alcohol intake and tobacco smoking were measured on a 5-point self-report scale, from 0 = none or not at all to 4 = heavily or very frequently. Alcohol intake was generally low, and few participants from either group smoked: alcohol-exposed group ($\bar{x}$ 1.76, SD 0.9); control group ($\bar{x}$ 1.78, SD 0.75); tobacco smoking-exposed group ($\bar{x}$ 0.32, SD 0.82); control group ($\bar{x}$ 0.83, SD 0.87). Most women reported they were in very good or excellent health in a 7-point scale, see Idler & Kasl [23]: exposed group ($\bar{x}$ 6.05, SD 0.83); control group ($\bar{x}$ 6.02, SD 0.81.7).

The screening survey included questions on injury that might confound the psychomotor test results, other environmental influences on health, and work history. For the exposed group there were additional questions on mercury hygiene practices and recall of mercury spill accidents.

The 2-hour neurobehavioural test battery was designed to be culturally appropriate for a cohort of New Zealand mid-life women. It contained a dual purpose memory and malingering test. Psychomotor tests included a computerised simple reaction time test, the O'Connor Tweezer Dexterity Test, the Grooved Pegboard, the Jamar dynamometer for grip strength, and the Tremorometer® for finger tremor.

3. Procedure

The assessments were conducted in English, in most case, in participants' homes. All the standardised tests had developers' instructions and these were followed.

For finger tremor the accelerator sensor was taped to the first finger for trials in three positions for each hand. Each trial lasted 21 seconds with 100 recordings per second. The Tremorometer measured frequency, amplitude and intensity of tremor on three-axes, and was supported by a computer software package to provide summary statistical data and Fast Fourier Transformed (FFT) power spectra. The hand positions tested related to tremors defined by Findley and Koller, p. 123 [24]. The first trial was for "postural tremor", a tremor that occurs when maintaining an action against gravity. The participant extended an arm in front, at shoulder height. The second trial was for "rest tremor", a tremor that occurs with the hand inactive, and in a hanging-relaxed position. The third trial was for "action tremor", and was measured as the voluntary action of the hand against gravity and holding a light load. In this case the participant repeated the posture position with an added 135 g weight.

Psychomotor data were analysed for between group differences using Statistical Package for Social Sciences (SPSS) v10, with Bonferroni adjustment to significance levels for multiple testing of the same participants. Tremor data were analysed with the TremorLab® analytical software package.

The project was assessed and approved by the Massey University Human Ethics Committee (MUHEC), protocol number PN 01/125.

4. Results

Participants were assigned to four levels of mercury exposure based on reported work history, with all controls in the no exposure level, and exposed women bands of up to five years exposure, five to10 years of exposure and more than 10 years. One participant reported traumatic brain injury and was excluded from analysis. Occupational overuse syndrome (OOS) was reported by 32.5% of exposed group and 06.7% of the control group. All participants from the exposed group reporting OOS had worked in dentistry more than 5 years. This was a significant between groups difference ($\chi^2 = 6.80$, (1, N = 70), $p < .01$). There was no evidence of malingering.

Simple reaction time For each participant, from 60 trials, reaction times in milliseconds were rank ordered and the slowest 10 and fastest 10 trials were excluded to minimise the effects of inattention and anticipation respectively. Comparing the central 40 trials by mean scores there was no significant difference between the exposed group ($\bar{x}$ = 292.78, SD = 32.43) and the control group ($\bar{x}$ = 292.14, SD = 29.05), t = 0.8, p = 0.93 (two tailed), df = 67.

O'Connor Tweezer Dexterity Test. Time in minutes and seconds for placement of 100 pins was converted to minutes and decimal fractions of a minute (min), and contributed to the group data. Levene's test for equal variances showed that data for exposed and control groups did have significantly different variance (F = 8.15, p = .006) with the control group scores negatively skewed. With the assumption of unequal variances there was no significant difference between the exposed group ($\bar{x}$ = 6.24 min, SD = 0.86 min) and the control group ($\bar{x}$ = 6.66 min, SD = 1.36 min), t = 1.46, p = .15 (two tailed), df = 46.55.

Grip strength For each participant, raw scores from three trials with each hand were averaged and scores recorded for dominant and non-dominant hands in kilograms force. Comparing the mean scores for the dominant hand, there was no significant difference between the exposed group ($\bar{x}$ = 28.96, SD = 6.79) and the control group ($\bar{x}$ = 28.88, SD = 4.24), t = 0.06, p = 0.95, (two tailed), df = 71. Comparing the mean scores for the non-dominant hand, there was no significant difference between the exposed group ($\bar{x}$ = 27.89, SD = 5.25) and the control group ($\bar{x}$ = 27.77, SD = 4.85), t = 0.10, p = .92, (two tailed), df = 71.

Grooved Pegboard Scores for the Grooved Pegboard were recorded in seconds and centi-seconds for time taken to place correctly all 25 pegs in the board. Comparing the mean scores, there was no significant difference between the exposed group ($\bar{x}$ = 69.01, SD = 9.26) and the control group ($\bar{x}$ = 68.28, SD = 10.75), t = 0.31, p = 0.76, (two tailed), df = 70.

Pearson's r test was used to correlate the mean scores on the Grooved Pegboard with the O'Connor Tweezer Dexterity Test. This showed that for the exposed group, the scores had a moderate positive correlation, r = +0.50, while the control group had a weaker positive correlation, r = +0.32. The difference was not significant.

Tremor Finger tremor patterns showed a significant, unilateral, negative correlation between duration of exposure and tremor mean frequency. All these data were collected with the one meter. Correlations between duration of exposure and three hand positions were: "right rest" r = −0.68, p = 0.03; "right posture" r = −0.78, p = 0.003; "right load" r = −0.74, p = 0.002. Duration correlations used exposed group data only as the control group had no exposure. There were no significant correlations between tremor frequency and duration of exposure in the left (non-dominant) hand. A second instrument was used during subsequent data collection but there was no ability to calibrate the two meters so this is not reported or discussed beyond **Table 2**.

With the final data set, a GLM ANOVA was conducted to examine the effect on tremor frequency, measured in Hz, from several independent variables: meter (two levels), position (three levels), hand (2 levels), and mercury exposure (four levels). None of the IVs are repeated measures as there was too much missing data from meter 2 to impute the values necessary for a mixed measure ANOVA. There was a significant main effect found for meter and for position but no significant interactions (see **Table 2**). All participants were right-hand dominant.

Table 2. Between-Groups ANOVA: mean tremor frequencies (Hz).

	Frequency Mean (SD)	df	F	*p*	η^2
Meter					
Meter 1	8.08 (0.99)	1.172	33.17	<0.001	0.16
Meter 2	6.84 (1.71)				
Position					
Posture	7.93 (1.63)	2.172	8.57	<0.001	0.09
Rest	6.78 (1.35)				
Load	7.53 (1.44)				
Hand					
Dominant	7.38 (1.62)	1.172	0.16	0.69. ns.	<0.01
Non Dominant	7.45 (1.48)				
Exposure					
None	7.61 (1.58)	3.172	0.91	0.44. ns.	0.02
Low	7.38 (1.48)				
Medium	7.32 (1.35)				
High	7.07 (1.66)				

A separate analysis with repeated measures for hand and for position was conducted on data from meter 1. There was a main effect for position and hand, with the right hand showing a significant decrease in frequency between the mean posture and rest positions, 8.8 Hz and 8.7 Hz respectively, and the load position mean of 7.4 Hz.

An in-depth analysis of the sinusoidal waveforms was conducted. The TremorLab provided graphs of the 20 second tremor samples by three axes, as a composite recording, and these could be broken into one second blocks and single cycles, as well as frequency and power spectra displays. From the visual data display, with exposed group data only, there was an apparent break or stall in the waveforms. **Figure 2** and **Figure 3** show tremor samples from the dominant

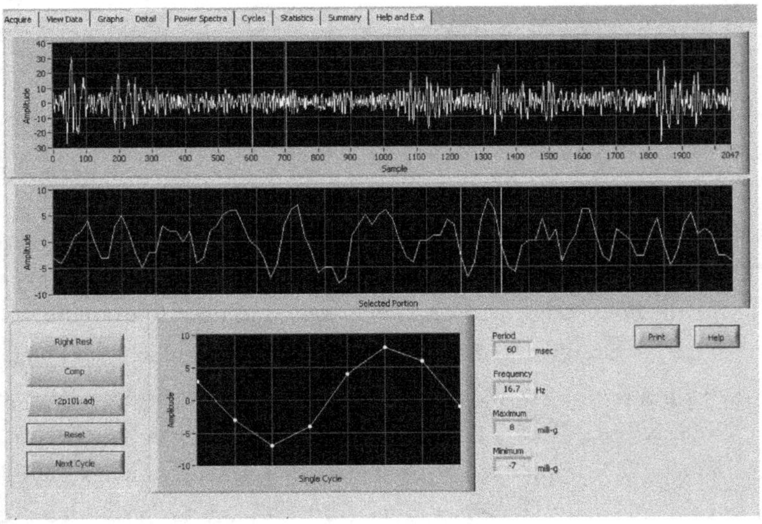

Figure 2. Control participant's 20 second composite tremor sample, right hand, rest position; with a one-second area selected for magnification (yellow lines), and a single "normal" cycle.

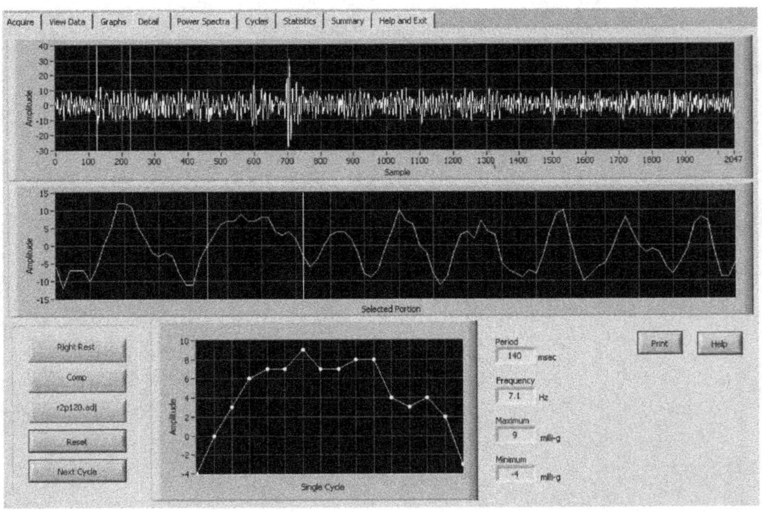

Figure 3. Exposed participant's 20 second composite tremor sample, right hand, rest position; with a one-second area selected for magnification (yellow lines), and a single cycle illustrating a "stalled" waveform.

hand in a rest position. **Figure 2** is a control participant, and **Figure 3** the mercury exposed participant. In both figures the top line shows the 20 second sample with yellow lines highlighting a one-second section selected to illustrate the differences; a middle line showing the highlighted one-second section displayed in detail, and beneath that a single cycle. The one-second display and one single cycle show a normal waveform in the control participant, and a stalled cycle in the mercury exposed participant

5. Discussion

Throughout the mercury poisoning literature, reports of damage to the peripheral nervous system are common, and the specific psychomotor tasks of grip strength, the Grooved Pegboard and simple reaction time were included for this reason. These assessment tools are well established for mercury exposure use [24] [25]. However no residual psychomotor skill differences were shown between the exposed group and controls, although a within group test of the exposed group reported significantly greater experience of OOS the longer the exposure period. While mercury may be implicated, other factors may be responsible such as the effects of years of holding small vibrating devices [26].

The inclusion of the O'Connor Tweezer Dexterity test was related to the exposed group's practice of dentistry and added an additional challenge to the skill required for the Grooved Pegboard. The exposed group had used tweezers for much of their detailed daily work, and some residual transfer of training may have occurred. Finger dexterity, assessed with the "pins-mins" test, differentiated mercury exposed and controls with practicing dentists [27] [28] [29], but even with the finer hand-eye coordination required with tweezers, no differences were found in the present study. This supports the assertion that there were no residual hand-eye coordination difficulties for the exposed group. Data were collected in 2003 and initially the no-difference finding meant the psychomotor results were not reported in other publications of the study; but the tremor analysis continued.

Tremor is the single most cited symptom of mercury poisoning yet there has been very little standardisation of visual analytical techniques or equipment. A Danish device that incorporated a two-axes accelerometer in the tip of a pen-like instrument had gained acceptance at the time of data collection, [30] but the Tremorometer [31], while largely untested, offered a three axis accelerometer, so was preferred . However differences in the two meters used, accounted for 16% of variance in tremor. Not having a single meter or being able to use them both with the same participant meant reliability was not assured, and this is a limitation of the study.

What the Tremorometer could do was show a frequency window for mercury-tremor, and tremor characteristics; and through the analytic software, had the potential for differential diagnosis of mercury tremor from essential tremor and Parkinsonism tremor, as described by others [32] [33] [34] when using an axis by axis, cycle by cycle view of a sample of tremor (see **Figure 2** and **Figure 3**). The three axis recordings could also be separated and compared with the power spectra images but at this point in the analysis there was nothing unexpected. However throughout the exposed group there was an apparent sinusoidal wave stall that, if investigated further, may be a mercury biomarker or "fingerprint". It is in this that the main contribution from the present study may be made. Since tremor testing is non-invasive and brief, the identification of such a fingerprint could be useful in monitoring people currently occupationally exposed to mercury for changes in individual tremor response at a sub-clinical level, as well as in group studies. The author invites comment on this point, particularly from researchers investigating other heavy metals.

6. Conclusion

In a 30-year follow-up study of mid-life women occupationally exposed to mercury, there was little evidence to show they had been compromised in psychomotor skill by their earlier exposure. The exception was in the self-reporting of OOS, which was strongly and positively correlated with length of time working with mercury. Sub-clinical dominant hand tremor was found in the exposed group women, but although mercury and tremor are known to be associated, in this sample, there were other factors that could not be ruled out. However from the tremor data the possibility of a unique mercury "fingerprint" was noted, and if this could be corroborated by other researchers, may prove highly valuable as a screening or monitoring tool in occupational health and safety where mercury is concerned.

References

[1] Appendices to the Journal of the House of Representatives (New Zealand) (1975) 33-34, 88-89.

[2] Appendices to the Journal of the House of Representatives (New Zealand) (1976) 31, 40, 90.

[3] Frykholm, K. (1957) Mercury from Dental Amalgam: Its Toxic and Allergic Effects. *Acta Odontologia Scandinavica*, **15**, 1-108.

[4] World Health Organisation, International Program on Chemical Safety. http://www.who.int/ipcs/en/ Accessed 08/14/2017

[5] Meyer-Baron, M., Schaeper, M. and Seeber, A. (2002) A Meta-Analysis for Neurobehavioural Results due to Occupational Mercury Exposure. *Archives of Toxicology*, **76**, 127-136. https://doi.org/10.1007/s00204-002-0327-9

[6] Short, A. (1974) Use of Mercury in Restorative Dentistry (Editorial Notes). *New Zealand School Dental Service Gazette*, **6**, 63-64.

[7] Jones, L. (2009) We Gave the Good Patients Mercury in Burr Boxes. A View of Mercury from Its Earlier Use in the School Dental Service to Contemporary Dental Practice. *New Zealand Dental Journal*, **105**, 47-50.

[8] Echeverria, D., Heyer, N., Martin, M., Naleway, C., Woods, J. and Bittner, A. (1995) Behavioural Effects of Low-Level Exposure to Hgo among Dentists. *Neurotoxicology and Teratology*, **17**, 161-168. https://doi.org/10.1016/0892-0362(94)00049-J

[9] Ngim, C., Foo, S., Boey, K. and Jeyaratnam, J. (1992) Chronic Neurobehavioural Effects of Elemental Mercury in Dentists. *British Journal of Industrial Medicine*, **49**, 782-790.

[10] Shapiro, I., Cornblath, D., Sumner, A., Uzzell, B., Spitz, L., Ship, I. and Bloch, P. (1982) Neurophysiological and Neuropsychological Function in Mercury-Exposed Dentists. *The Lancet*, **319**, 1147-1150. https://doi.org/10.1016/S0140-6736(82)92226-7

[11] Uzzell, B. and Oler, J. (1986) Chronic Low-Level Mercury Exposure and Neuropsychological Functioning. *Journal of Clinical and Experimental Neuropsychology*, **8**, 581-593.

[12] Iyer, K., Goodgold, J., Eberstein, A. and Berg, P. (1976) Mercury Poisoning in a Dentist. *Archives of Neurology*, **33**, 788-790. https://doi.org/10.1001/archneur.1976.00500110056011

[13] Albers, J., Kallenbach, L., Fine, L., Langolf, G., Wolfe, R., Donofrio, P., *et al.* (1988) Neurological Abnormalities Associated with Remote Occupational Elemental Mercury Exposure. *Annals of Neurology*, **24**, 651-659. https://doi.org/10.1002/ana.410240510

[14] Letz, R., Gerr, F., Cragle, D., Green, R., Watkins, J. and Fidler, A. (2000) Residual Neurologic Deficits 30 Years after Occupational Exposure to Elemental Mercury. *Neurotoxicology*. **21**, 459-474.

[15] Frumkin, H., Letz, R., Williams, P., Gerr, F., Pierce, M., Sanders, A., *et al.* (2001) Health Effects of Long-Term Mercury Exposure among Chloralkali Plant Workers. *American Journal of Industrial Medicine*, **39**, 1-18. https://doi.org/10.1002/1097-0274(200101)39:1<1::AID-AJIM1>3.0.CO;2-N

[16] Powell, T. (2000) Chronic Neurobehavioural Effects of Mercury Poisoning on a Group of Zulu Chemical Workers. *Brain Injury*, **14**, 797-814. https://doi.org/10.1080/026990500421912

[17] Ratcliffe, H., Swanson, G. and Fischer, L. (1996) Human Exposure to Mercury: A Critical Assessment of the Evidence of Adverse Health Effects. *Journal of Toxicology & Environmental Health*, **49**, 221-270. https://doi.org/10.1080/00984108.1996.11667600

[18] Cavalleri, A. and Gobba, F. (1998) Reversible Colour Vision Loss in Occupational Exposure to Metallic Mercury. *Environmental Research*, **77**, 173-177. https://doi.org/10.1006/enrs.1997.3814

[19] Triebig, G. and Schaller, K. (1982) Neurotoxic Effects in Mercury-Exposed Workers. *Neurobehavioral Toxicology & Teratology*, **4**, 717-720.

[20] Wood, R., Weiss, A. and Weiss, B. (1973) Hand Tremor Induced by Industrial Exposure to Inorganic Mercury. *Archives of Environmental Health*, **26**, 249-252. https://doi.org/10.1080/00039896.1973.10666268

[21] Jones, L., Bunnell, J. and Stillman, J. (2007) A 30-Year Follow-Up Study of Women's Occupational Mercury Exposure: An Analysis of Residual Effects. *Human and Experimental Toxicology*, **26**, 367-374. https://doi.org/10.1177/0960327107076824

[22] Ambler, R., Anger, K. and Sizemore, O. (1995) Adult Environmental Neurobehavioural Test Battery. Department of Health and Human Services, Agency for Toxic Substances and Disease Registry, Atlanta.

[23] Idler, E. and Kasl, S. (1991) Health Perceptions and Survival: Do Global Predictions of Health Status Really Predict Mortality? *Journal of Gerontology*, 46, 55-65. https://doi.org/10.1093/geronj/46.2.S55

[24] Findley, L. (1996) Classification of Tremors. *Journal of Clinical Neurophysiology*, **13**, 122-132. https://doi.org/10.1097/00004691-199603000-00003

[25] Hartman, D. (1995) Neuropsychological Toxicology: Identification and Assessment of Human Neurotoxic Syndromes. 2nd Edition, Plenum Press, New York. https://doi.org/10.1007/978-1-4615-1849-5

[26] Heaver, C., Goonetilleke, K., Ferguson, H. and Shiralkar, S. (2011) Hand-Arm Vibration Syndrome: A Common Occupational Hazard in Industrialized Countries. *Journal of Hand Surgery* (*Eur*), **36**, 354-363. https://doi.org/10.1177/1753193410396636

[27] Bittner, A., Echeverria, D., Woods, J., Aposhian, H., Naleway, C., Martin, M., *et al.* (1998) Behavioral Effects of Low-Level Exposure to Hg0 among Dental Professionals: A Cross-Study Evaluation of Psychomotor Effects. *Neurotoxicology & Teratology*, **20**, 429-439. https://doi.org/10.1016/S0892-0362(98)00006-3

[28] Echeverria, D., Aposhian, H., Woods, J., Heyer, N., Aposhian, M., Bittner, A., *et al.* (1998) Neurobehavioural Effects from Exposure to Dental Amalgam Hgo: New Distinctions between Recent Exposure and Hg Body Burden. *FASEB Journal*, **12**, 971-980.

[29] Gonzalez-Ramirez, D., Maiorino, R., Zuniga-Charles, M., Xu, Z., Hurlbut, K., Junco-Munoz, P., *et al.* (1995) Sodium 2,3-Dimercaptopropane-1-sulfonate Challenge Test for Mercury in Humans: Urinary Mercury, Porphyrins and Neurobehavioral Changes of Dental Workers in Monterrey, Mexico. *Journal of Pharmacology & Experimental Therapeutics*, **272**, 264-274.

[30] Galinski, T., Rosa, R. and Wheeler, D. (1990) Assessing Muscular Fatigue with a Portable Tremor Measurement System Suitable for Field Use. *Behavioral Research Methods, Instruments, and Computers*, **22**, 507-516. https://doi.org/10.3758/BF03204434

[31] Tripp, R. (2002) Tremerometer User Manual. FlexAble Systems Inc., Fountain Hills, AZ.

[32] Beuter, A and de Geoffrey, A. (1996) Can Tremor Be Used to Measure the Effects of Chronic Mercury Exposure in Human Subjects? *Neurotoxicology*, **17**, 213-228.

[33] McCullough, J., Dick, R. and Rutchik, J. (2001) Chronic Mercury Exposure Examined with a Computer-Based Tremor System. *Journal of Occupational & Environmental Medicine*, **43**, 295-300. https://doi.org/10.1097/00043764-200103000-00022

[34] Biernat, H., Ellias, S., Wermuth, L., Cleary, D., de Oliveira Santos, E., Jorgensen, P., *et al.* (1999) Tremor Frequency Patterns in Mercury Vapor Exposure, Compared with Early Parkinson's Disease and Essential Tremor. *Neurotoxicology*, **20**, 945-952.

12

Air Quality, Health and Community Action

Larry E. Erickson[1], Wendy Griswold[2], Ronaldo G. Maghirang[1], Brian P. Urbaszewski[3]

[1]Kansas State University, Manhattan, KS, USA
[2]University of Memphis, Memphis, TN, USA
[3]Respiratory Health Association, Chicago, IL, USA

Email: lerick@ksu.edu, wgrswold@memphis.edu, rmaghir@ksu.edu, BUrbaszewski@lungchicago.org

Abstract

Air quality is impacting health in many cities in most countries because of particulate pollution, sulfur dioxide, nitrogen dioxide, and ozone. Very small particulates from engine emissions and coal fired electric power plants enter the lungs and pollute the blood of urban residents leading to a burden of disease with more than 3 million premature deaths per year attributed to outdoor air pollution. Welfare losses including premature deaths associated with air pollution were about $5 trillion in 2013. A global transition to electric vehicles, and the generation of electricity without combustion emissions would improve air quality significantly and reduce greenhouse gas emissions. This transition is in progress in many parts of the world with more than 2 million electric vehicles in service in 2017. Electric bus and electric taxi sales are increasing, and many large cities have multiple programs to improve air quality. When health costs are considered, it is very appropriate for communities to take action to improve air quality and health. This work reviews and reports many positive actions that are in progress in larger cities.

Keywords

Air Pollution, Particulates, Air Quality, Health, Electric Vehicles, Solar, Disease, Cities, Urban

1. Introduction

Reducing greenhouse gas emissions and improving air quality are significant global challenges. This review addresses air quality, the impact of air pollution on health, and community actions to improve air quality and associated health equity [1]-[98].

Air pollution is a global concern with social costs of the order of $3 trillion/year [21] [22] [43] [51] [58]. There are about 3.3 million premature deaths per year due to outdoor air pollution, and particulate matter (PM2.5) is one of the most important air pollutants [48]. Most major urban centers have concentrations of particulates that are above the recommended World Health Organization Air Quality Guideline of 10 micrograms per cubic meter (μg/m^3) [21] [22] [94]. Reducing PM2.5 concentrations to less than the WHO guidelines would prevent an estimated 2.1 million deaths per year [4] [80]. The health care expenses associated with poor air quality would be reduced significantly if urban air quality were improved to meet regulatory values [80]. A relatively new World Bank/Institute for Health Metrics and Evaluation study estimates that the welfare losses due to deaths from air pollution were about $5 trillion in 2013 [80] [96]. The American Lung Association of California has reported that air pollution attributed to passenger vehicles had an estimated harm of $24 billion for health and $13 billion for climate in 2015 in the states of California, Connecticut, Maine, Maryland, Massachusetts, New Jersey, New York, Oregon, Rhode Island, and Vermont [38]. This group of states is working cooperatively to encourage electric vehicle adoption and reductions in vehicle emissions. In the United States, there are about 200,000 premature mortalities per year with emissions from vehicles and coal burning power plants being the most significant sources [15]. Air pollution from burning coal includes small particulates containing sulfur and nitrogen compounds. The largest source of sulfur emissions is coal combustion [15].

One purpose of this manuscript is to help residents of urban communities in their efforts to understand the air quality where they live, work, and play. Many residents of large cities do not realize that particulates in air have been reported to be the fifth largest overall global risk factor for premature mortality [34]. Another purpose of this manuscript is to provide information on the health effects of air pollution. The third goal of the manuscript is to include some information on actions that can be taken to improve air quality in urban communities.

Urban cities and metropolitan areas can have air quality management plans (AQMP), which include air quality objectives, monitoring, emission inventory, prediction tools, control strategies, and public participation [32] [81]. Improvements in air quality can be achieved by implementation of the AQMP to accomplish goals that are important for the community.

2. Air Quality in Urban Communities

Particulates (PM2.5), nitrogen dioxide (NO_2), ozone (O_3), carbon monoxide (CO), and sulfur dioxide (SO_2) are the most common air pollutants in urban communities [23] [85] [87]. These pollutants are included in the air quality index (AQI) that has been developed to inform residents about air quality in their communities [85]. The index has 6 categories with a color for each: good, moderate, unhealthy for sensitive groups, unhealthy, very unhealthy, and hazardous [85]. The corresponding colors are green, yellow, orange, red, purple, and maroon. The U.S. EPA has established National Ambient Air Quality Standards

(NAAQS) for each criteria pollutant [87]. The NAAQS for NO_2, O_3, and particulates are shown in **Table 1**. In **Table 1**, the primary standards are to protect public health, including that of more sensitive populations such as asthmatics. Secondary standards are to provide public welfare protection, including damage to plants and animals [87]. The AQI for NO_2, O_3, and particulates is shown in **Table 2**. The European Commission of the European Union has developed air quality standards for the countries in the European Union [20]. The air quality standards are designed to protect human health and public welfare (*i.e.*, effects on vegetation, weather, visibility, etc.).

Table 1. National Ambient Air Quality Standards for PM2.5, nitrogen dioxide (NO_2), and ozone (O_3) [87].

Pollutant		Primary/ Secondary	Averaging Time	Level	Form
Nitrogen Dioxide (NO_2)		Primary	1 hour	100 ppb	98th percentile of 1-hour daily maximum concentrations, averaged over 3 years
		Primary and Secondary	1 year	53 ppb	Annual Mean
Ozone (O_3)		Primary and Secondary	8 hours	0.070 ppm	Annual fourth-highest daily maximum 8-hour concentration, averaged over 3 years
Particle Matter	$PM_{2.5}$	Primary	1 year	12.0 μg/m^3	Annual mean, averaged over 3 years
		Primary and Secondary	24 hours	35 μg/m^3	98th percentile, averaged over 3 years
	PM_{10}	Primary and Secondary	24 hours	150 μg/m^3	Not to be exceeded more than once per year on average over 3 years

Table 2. Air quality index (AQI) [87].

AQI Categories	Health Concern	Ozone (ppm)	PM2.5 (μg/m^3)	NO_2 (ppb)
Good (up to 50)	Air quality is satisfactory	0 - 0.054	0 - 12.0	0 - 53
Moderate (51-100)	Air quality is acceptable; however, unusually sensitive people should consider reducing prolonged or heavy exertion.	0.055 - 0.070	12.1 - 35.4	54 - 100
Unhealthy for Sensitive Groups (101 - 150)	People with heart and lung disease, older adults and children are at a greater risk from exposure to ozone and particulate matter.	0.071 - 0.085	35.5 - 55.4	101 - 360
Unhealthy (151 - 200)	Everyone may begin to experience some adverse health effects, and those in the sensitive groups may experience more serious health effects.	0.086 - 0.105	55.5 - 150.4	361 - 649
Very Unhealthy (201 - 300)	Everyone may experience more serious health effects	0.106 - 0.200	150.5 - 250.4	650 - 1249
Hazardous (301 - 500)	Entire population is more likely to be affected.	-	250.5 - 500.4	1250 - 2049

2.1. Particulate Matter in Air

Particulate matter in air is one of the most important air pollutants. The U.S. EPA NAAQS for particulates that are 2.5 micrometers (μm) in equivalent aerodynamic diameter or smaller (PM2.5) are 12 $\mu g/m^3$ (annual average) and 35 $\mu g/m^3$ (24-hour average) [87]. The air quality index (AQI) for PM2.5 has a value of 50 when PM2.5 is equal to 12 for the transition value from green to yellow and a value of 100 when PM2.5 is equal to 35.4 for the transition value from yellow to orange. The AQI value of 100 corresponds to the 24 hour PM2.5 NAAQS value of 35 $\mu g/m^3$ [85].

Particulate matter impacts on health depend on the size distribution of the particulates and the chemicals that make up the particulates. The size of particulates from internal combustion vehicle emissions are very small compared to dust from soil. Particulates associated with vehicle emissions range from about 10 nanometers (nm) to 700 nm [44] [46] [97]. Because of their small size they pass into the smaller parts of the lung and some find their way into the blood stream and the brain [10]. Particulates with equivalent aerodynamic diameters in the range of 5 to 100 nm are defined as ultrafine particles [12]. Carbon compounds from diesel exhaust in particulates have been identified with increases in the number of cases of lung cancer [2] [10] [22] [40].

Concentrations of particulate matter in major cities in micrograms per cubic meter are well above the WHO recommended value of 10 $\mu g/m^3$; annual average values include 170 for Delhi, 66 for Cairo, 55 for Beijing, 19 for Los Angeles, and 13 $\mu g/m^3$ for Chicago [22] [45]. Values of particulate matter are strongly associated with vehicle emissions and traffic density. Concentrations are larger near freeways and roads with many vehicles [21].

2.2. Nitrogen Dioxide in Air

The regulatory concentrations of nitrogen dioxide in ambient air are 100 parts per billion (ppb) or 188 $\mu g/m^3$ (1 hour average value) and 53 ppb or 100 $\mu g/m^3$ (annual mean value) [87]. The Air Quality Index has the good range (AQI from 0 to 50) from 0 to 53 ppb and the moderate range (AQI from 51 to 100) from 54 to 100 ppb for nitrogen dioxide [85].

About 58% of nitrogen oxides in urban air are attributed to mobile sources [23]. Other combustion sources also have emissions of nitrogen oxides. Since nitrogen dioxide emissions are associated with vehicle emissions, the concentrations in urban air vary with time of day and traffic density.

2.3. Ozone in Air

Ozone is produced in urban air by photochemical reactions when hydrocarbons and nitrogen oxides are present from vehicle emissions and other sources [23]. The regulatory concentration for ozone is 0.070 parts per million (ppm) or 70 ppb. This is an 8 hour average value. The regulation is written such that this average value may happen on up to three days per year averaged over three years

[87]. The Air Quality Index has the good range (AQI from 0 to 50) from 0 to 0.054 ppm and the moderate range (AQI from 51 to 100) from 0.055 to 0.07 ppm [35].

2.4. Traffic Density and Air Quality

Air quality is impacted by stationary point sources such as coal fired power plants industrial smoke stacks and mobile sources such as vehicles. Since traffic density varies with location and time of day, there will be variations in air quality that depend on how close vehicle emissions are to the instruments that are making air quality measurements. There are several studies that report air quality data near roads and streets [7] [11] [12] [63]. Particulate numbers and concentrations are significantly larger near major freeways and highways, and values vary with traffic density. There is often a concentration gradient over a distance of about 100 to 300 meters with much smaller particulate concentrations 100 meters away from the road [12]. Air quality parameters measured near roads are commonly significantly larger than those measured at fixed regulatory monitoring sites; values twice as large near roads have been reported [11].

3. Health Effects of Air Pollution

Since air pollution includes particulates, sulfur dioxide, nitrogen oxides and ozone epidemiology studies are impacted by more than one pollutant in many cases It is well known that these pollutants are causally related to cardiovascular disease [10]. Individuals with asthma and/or chronic obstructive pulmonary disease may be impacted by these pollutants in the air they breathe. Lung cancer mortality is associated with particulates from diesel engines [2] [10] [22].

There is evidence that compounds in urban air pollutants move from the lungs to the blood stream and the brain [10, 29]. Decreased cognitive function, autism in children, Parkinson's disease, and dementia are identified as being more prevalent in large cities because of air pollution [10] [14] [21] [61] [79]. Traffic related studies have reported that dementia is associated with air pollution from vehicles including small particulates [18]. The association of Parkinson's disease with air pollution appears to be not as strong as with dementia [10] [18] [29] [68]. About 47 million people worldwide are living with dementia, and the estimated annual cost is $818 billion [65].

There are several studies that report an increase in the percent of children with autism because of increased concentrations of air pollutants [6] [66] [69] [90]. The associations were stronger for boys than girls [69]. The third trimester of pregnancy appears to have the strongest association with the health impact of the concentration of air pollutants on autism [66]. Living near a freeway was associated with autism [90]. Traffic sources of air pollution were significant in the study by Becerra *et al.* [6].

There are significant effects of air pollution due to PM2.5 and ozone on adverse birth outcomes [27] [33] [55]. The risks associated with low birth weight, preterm delivery, and very preterm delivery are greater for babies of women who are exposed daily to high concentrations of PM2.5 while pregnant [33]. Ozone appears to increase the risk of early delivery, but it decreases the risk of term low birth weight [33]. The estimated annual cost of preterm births associated with air pollution in the United States has been estimated to be about $5 billion with an estimated 3.32 % of the preterm births attributed to PM2.5 being greater than 8.8 micrograms per cubic meter [77]. Sulfur dioxide contributed to low birth weight babies in China [91].

Childhood leukemia appears to have an association with air pollution. There may be exposures to air pollution during pregnancy and after birth. Traffic-related air pollution studies have shown an association of traffic density with childhood leukemia [26] [30] [35].

Multiple sclerosis (MS) is a neurological disorder and chronic inflammatory disease of the central nervous system that has been associated with air pollution [10] [29] [57]. Research has found a relationship between air pollutants including PM and MS relapses [29] [36] [53] [60] [89]. Cigarette smoking and smoke have also been identified as exposures with associated health impacts for those who have MS [59].

Recently, the literature related to air pollution and autoimmune disease development has been reviewed for rheumatoid arthritis (RA), juvenile idiopathic arthritis (JIA), and systematic autoimmune rheumatic diseases (SARDs) [75]. An association between PM2.5 air pollution and SARDs was found [8] [9]. For children less than 5.5 years, an association was found for JIA and PM2.5 [98]. Further research is suggested on the effect of air pollution on autoimmune diseases [75].

Data from the Global Burden of Disease Study 2013 has been used to estimate the stroke-related disability adjusted life-years (DALYs) associated with a number of risk factors worldwide from 1990 to 2013 [25]. Globally, 29.2% of the burden of stroke was attributed to air pollution [25]; this included both ambient air pollution and household air pollution from solid fuels. Almost 22% of stroke related DALYs were attributed to ambient PM2.5 air pollution in China and India compared to 10.2% in high income countries [25]. The total number of DALYs in 2013 was about 102 million for stroke; more than 80% of these were from low-income and middle-income countries [25].

The health impacts associated with air pollution may be difficult to measure because of other factors that also affect health such as level of education, psychosocial stressors and quality of medical care [56]. Air quality varies within cities and low income communities often have higher air pollutant concentrations in their neighborhoods. There is no safe level of exposure to some air pollutants [67].

4. Improving Air Quality

There are a number of actions that can be taken to improve urban air quality.

Since coal combustion is an important source of air pollutants, the transition from coal to wind and solar energy for generation of electricity is a high priority. A number of communities have programs in place with a goal to reduce emissions and air pollutant concentrations. In California, there are ongoing efforts to advance the electrification of transportation with incentives for residents to purchase electric vehicles and programs to encourage solar generation of electricity [16] [21] [22] [23]. California has an extensive electric vehicle charging infrastructure, including shaded parking where solar panels are producing electricity in parking lots that include charging equipment [23].

The global issue of greenhouse gas emissions and climate change would benefit if all large cities worked to improve their air quality by transitioning to electric powered transportation and if the electricity was generated without combustion using wind and solar energy, hydro and nuclear power. Some progress is being made as described below.

The U.S. EPA has a research program at the nexus of reducing greenhouse gas emissions and improving air quality [82] [83]. Research to prevent and reduce air emissions in urban areas is one of the goals of the strategic plan, which addresses the challenge of protecting health and the environment from the impacts of climate change and air pollution.

4.1 Electric Buses

There are several companies that are making electric buses for use in cities, and the buses are economical for cities to purchase and use [22]. New Flyer Industries, Proterra, Volvo, and BYD are in competition to expand their sales of electric buses. Proterra is a new U.S. company that has met California's zero emission bus rules [22]. BYD is a Chinese company that has plants in Lancaster, California and Komarom, Hungary. When the health benefits are considered, electric buses are very logical choices for large cities that are working to improve their air quality.

There are two approaches to charging the batteries on electric buses. One is to have a sufficient charge on the battery to only need to charge them each evening. The other alternative is to have wireless charging at locations where the bus stops along the route. One of the important efforts that is being addressed is the need for international standards for charging buses [41]. Solar panels can be installed at bus stops to provide shade and cover for those who are waiting for the bus and to generate electric power that can be stored in batteries at charging locations along the bus route.

Chicago Transit Authority purchased two electric buses from New Flyer Industries and placed them in service in October 2014 [19] [22]. The buses have 300 kWh Li-ion batteries and a safety and performance battery management system. The batteries can be charged each evening in 3 to 5 hours. The estimated service life is 12 years. The buses have a modern electric power steering system which is expected to reduce maintenance and repairs [19]. The estimated health benefits per bus per year are estimated to be about $55,000 because of the reduced emissions of particulates, and nitrogen oxides [19]. The buses have performed well and the Chicago Transit Authority is planning to purchase additional electric buses.

London has made progress in its efforts to electrify transportation and reduce air pollution. This includes a transition to electric buses with 51 added in September 2016, and a plan to continue to bring additional electric buses into service until all 300 single-deck buses are electric buses and 3100 double-deckers are hybrid-electric buses [41] [50] [76]. When these 51 new buses are added to the existing 22 all-electric buses, there are 73 electric buses with the 22 serving outer London [76]. The motivation for the transition to electric buses is because of the need to improve air quality. Reductions in particulate matter and nitrogen oxides in air are the two pollutants of greatest concern [76].

There has been positive progress in many other communities to transition to electric buses. In China, sales of new electric buses were more than 100,000 in 2016 [5]. Seattle, Washington has plans to purchase up to 120 all electric buses, with initial orders from several companies to gather information on the relative quality of the different buses that are available [47]. Reduced emissions, better air quality, and quieter ride have been identified as positive features of the electric buses [22].

4.2. Electric Taxi Cabs

One of the ways to reduce emissions and improve air quality is to transition to electric taxi cabs. A new factory designed to produce 20,000 electric taxi cabs per year for London and other cities is being built in central England [28]. The plan to transition to electric taxi cabs is supported by some government funding because of the desire to improve air quality in London and other cities. London has a daily fee for motor vehicles that enter the congested core of the city during weekday business hours. This fee is to reduce congestion, and it has higher fees for polluting vehicles in order to improve air quality [88]. Fees on vehicles associated with air quality goals can be used as an incentive to encourage electric taxi cabs as well as other electric vehicles.

There is a plan in Beijing, China to convert all of the taxi cabs to electric vehicles. All new taxi cabs are to be electric vehicles with the estimated cost of 67,000 new cabs being about $1.3 billion [49]. BYD has been testing electric taxis in several large cities in China. Air quality issues are one of the reasons for this effort. The national government has a new energy vehicle campaign to encourage both plug-in hybrid and electric vehicle sales [49].

Electric vehicles are being used for taxi cabs in Montreal, Canada, New York City, Austin Texas, Amsterdam, Budapest, Barcelona, Prague, Rome, Stockholm, and many other places [52] [78]. Many cities are attempting to improve air quality by encouraging electric taxi cabs.

As the technology for wireless charging develops, it will be possible to have wireless charging while taxi cabs are waiting for customers at locations where there is frequent demand for passengers. Solar panels over these locations can be used to provide shaded charging and electricity. There can be energy storage at these locations also.

Newer ride share services such as Uber and Lyft have the potential to transition to electric vehicles and help with efforts to improve air quality and their image.

4.3 Other Ways to Improve Air Quality

There are many additional actions that can be taken to improve air quality and reduce emissions in communities. One approach is to reduce vehicle emissions by promoting walking and bicycle riding. Good sidewalks, bike lanes, and bike paths away from traffic encourage increased walking and cycling. Parking for bicycles is also important. The integration of biking and public transportation to allow bicycles on public transport systems is helpful. Electric bicycles should be promoted by having covered parking and charge stations. Bicycle share programs in which bicycles are available for riding from one place to another can be established in communities [37].

Low emission zones in cities where vehicles that do not meet emission standards are restricted from entering have been established in many cities [37] [92]. Car-free zones have been established in many cities as well.

One significant approach to improving urban air quality is to increase the amount of vegetation in urban areas along streets, in parks, with green roofs, and on balconies [54] [71]. Vegetation provides surfaces for particulates and gases to adsorb (stick to the surface of) and absorb (soak into) [42] [72]. Parks provide areas away from traffic emissions for people to exercise [73]. The literature related to the air quality benefits of green infrastructure in cities has been reviewed by Abhijith *et al.* [1]. Hedges along roads are recommended as a green solution that allows for the adsorption of pollutants to vegetation and the dispersion of the air.

Los Angeles, California has been working to reduce air pollution for about 50 years [62]. Progress toward better air quality has been made by mandating reductions in nitrogen oxide and sulfur oxide emissions from refineries, promoting electric vehicles, providing rideshare incentives, developing a clean up green up program to add vegetation, to get all electrical energy from renewable sources, to implement a clean air action plan to reduce air emissions at the Port of Los Angeles and the Port of Long Beach, to accurately identify sources of air pollution, and invest in strategies and technologies with multiple objectives regarding air quality air toxics, transportation, energy, and climate change [62]. In July 2017 the Metropolitan Transportation Authority decided to purchase 95 electric buses for use in Los Angeles County [99].

Public education on the health impacts of air pollution and actions that can be taken is recommended because many people do not understand the impact of air pollution on health [39]. Greater health awareness has the potential to change behavior and reduce vehicle emissions. Investment in cleaner vehicles and electric vehicles, reducing diesel idling, working from home, public transport subsidization, integration of government policies, and effective community participation in decision making have been identified as being beneficial in efforts to improve air quality [22] [39]. Recent developments in low cost air sensors make community monitoring a more affordable activity for educating residents about local air quality issues [95]. Increased knowledge and awareness of local health impacts and air quality has the potential to support meaningful participation in decision making at the community level and beyond. Several communities in the U.S. are engaged in community-based participatory research to determine how effective these new technologies are in supporting community learning and action related to air quality improvements [86].

The Air Quality Index (AQI) is used throughout the world to provide public information on air quality. Values from continuous monitoring are available on the Internet to inform the public about air quality in many locations [3] [93]. In some locations LED (light emitting diode) displays provide information on local air quality by reporting current local values of the AQI [70].

Local air quality can be included in the set of information that is used in making decisions. Senate Bill 352 in the State of California considers air quality in land-use decisions for new schools by not having schools along major roads [31]. Communities can include health considerations associated with air quality in community planning.

Personal behavior can impact the extent of exposure to air pollution associated with transportation. Some things individuals can do include avoiding rush hour traffic, checking air quality for current and projected readings, exercise in parks rather than along busy roads, consider air quality when deciding where to live, consider air quality and traffic density when making travel plans, filter the air in your home, keep windows closed if the outdoor air is poor, consider air quality for where your children play [73].

There are four recent publications that address actions that communities and/or governments can take to improve air quality and health [13] [17] [39] [92]. A tool has been developed to facilitate the assessment of health impacts of outdoor air pollution interventions by non-public health experts. Six categories of modifying factors have been identified: sources of emissions, quantity of emissions, concentrations of emitted pollutants, spatial distribution, personal exposure, and individual vulnerability [17]. This tool is intended to help municipal decision makers and program planners make better decisions as they design and evaluate proposed interventions to improve air quality.

In the process of searching for good practices to improve urban air quality, it is valuable to understand the community where new practices are to be implemented. Shandas *et al.* [74] have made use of spatial analysis to understand the community structure, population distribution, and air quality at the neighborhood scale. They have then selected technologies to improve air quality such as planting trees to enhance local air quality by adsorbing particulates.

A systematic review of 15 recent publications resulted in four strategies to improve urban air quality: general regulations on air emissions, road traffic related interventions, energy generation related interventions, and greenhouse gas emission reduction interventions [92]. Particulate matter and nitrogen dioxide were the pollutants of greatest concern in these 15 papers. Regulations and guidelines to reduce PM2.5 have resulted in improved air quality. The establishment of low emission zones and congestion charging schemes have been successful in reducing PM2.5 and nitrogen dioxide concentrations in the areas of the cities where they have been implemented. More than 200 low emission zones have been implemented in Europe [92]. Regulations that encourage electric vehicle purchase and use in cities improve air quality and reduce greenhouse gas emissions.

In homes and buildings, high efficiency filters can be used and better air intake locations for large buildings can be selected. Air intakes on roof tops or on sides of the building away from traffic should be considered [13].

Better understanding of the impact of air pollution on health in urban communities is identified as being important [39]. Better understanding can lead to better alignment of policies and measures to reduce air pollution with protecting and improving health. One potential action is to improve the quality and coordination of public transport services so they are more convenient and cost effective [39].

Another important community action that may have great impact is to promote a transition to electric vehicles (EVs) by providing incentives to purchase EVs, adding solar powered charging infrastructure in parking lots, and holding community education events that help residents understand the health benefits of reducing emissions. Adding solar panels and charging stations in parking lots has immediate benefits including clean electricity generation, shaded parking and charging infrastructure for EVs [23]. The shaded parking is beneficial because electric vehicle charging in the shade is better for the batteries in hot environments. Because of the decrease in costs of solar panels and batteries for EVs, there is the potential for a rapid transition to communities with a significant fraction of cars being EVs and many parking lots with solar powered charging stations [22]. The solar generated electricity and the car batteries in EVs can be mutually beneficial with the batteries being used as a repository for electric power that needs to be stored and the solar power providing clean power for EVs. Real time prices for electricity that encourage battery charging when prices are low are needed to encourage EV owners to charge their batteries at the best time [23].

The transition to electric vehicles may be encouraged by government policy. The United Kingdom, France, India, and Norway have announced plans to end the sale of new gasoline and diesel automobiles at a future date [24] [64]. Norway plans for all new passenger cars and vans sold in 2025 to be zero emission vehicles. In India every vehicle sold in the country in 2030 should be powered by electricity. France plans to end the sale of gasoline and diesel vehicles in 2040 [24]. The United Kingdom plans to end the sale of new gasoline and diesel cars in 2040 [64].

In the United States, the U.S. EPA has assisted with community efforts to improve air quality through the Community Air Risk Reduction Initiative (CARRI), the Community Action for a Renewed Environment (CARE) program, and the National Clean Diesel Campaign [83].

5. Conclusions

The particulates from vehicle emissions are very small with dimensions from about 10 to 700 nm and they are able to enter the lungs. Some compounds in the particulates enter the blood and are transported to other parts of the body. Air pollution associated with vehicle emissions is a very significant global problem, and annual social costs are of the order of $3 trillion with more than 3 million premature deaths per year associated with outdoor air pollution. The burden of disease on health that is partially attributed to air pollution includes cardiovascular disease, asthma, stroke, chronic obstructive pulmonary disease, lung cancer, dementia, multiple sclerosis, Parkinson's disease, autism in children, and childhood leukemia. Risks associated with low birth weight and early preterm delivery are greater for those living with higher levels of air pollution.

There are a number of actions that can be taken to improve air quality and health. The electrification of transportation and the transition to the generation of electricity with wind and solar energy are having significant impacts on improving air quality in California and some other parts of the world. Air pollution from coal combustion and industrial sources needs to be reduced in many locations. Electric buses are being purchased and put into service in the United States, Europe, and China. Electric taxi cabs provide another pathway to better air quality and health, and there is progress in their production and sales. More than 200 low emission zones have been established to improve air quality in regions of cities. Fees have been established and implemented to reduce congestion and vehicle emissions in some large cities.

Public education related to air pollution and health is important, and it is part of the process to improve health. Individuals can take action to reduce their exposure to air pollution including monitoring the air quality index, avoiding rush hour exposures, exercising in parks where particulate concentrations are lower, and filtering the air in their homes. Communities can take action to improve the quality of their air through public transport subsidization, promoting electric vehicles, and enhancing charging infrastructure for electric vehicles.

Acknowledgements

This publication was developed under Assistance Agreement RD83618201 awarded by the U.S. Environmental Protection Agency to Kansas State University. It has not been formally reviewed by EPA. The views expressed in this document are solely those of the authors and do not necessarily reflect those of the Agency. EPA does not endorse any products or services mentioned in this publication.

References

[1] Abhijith, K.V., Kumar, P., Gallagher, J., McNabola, A. *et al.* (2017). Air Pollution Abatement Performances of Green Infrastructure in Open Road and Built-Up Street Canyon Environments. Atmospheric Environment, **162**, 71-86. https://doi.org/10.1016/j.atmosenv.2017.05.014

[2] ACS (2016) Diesel Exhaust and Cancer, American Cancer Society Report. https://www.cancer.org/

[3] AirNow (2017) Air Quality Index—A Guide to Air Quality and Your Health, AirNow. https://www.airnow.gov/

[4] Apte, J.S., Marshall, J.D., Cohen, A.J. and Bauer, M. (2015). Addressing Global Mortality from Ambient PM2.5. *Environmental Science and Technology*, **49**, 8057-8066.

[5] Ayre, J. (2017) China 100% Electric Bus Sales Grew to about 115,700 in 2016, Clean Technica. https://cleantechnica.com/

[6] Becerra, T.A., Wilhelm, M., Olsen, J., Cockburn, M., and Ritz, B. (2013) Ambient Air Pollution and Autism in Los Angeles County, California. *Environmental Health Perspectives*, **121**, 380-386.

[7] Bereitschaft, B. (2015). Pedestrian Exposure to Near-Roadway PM2.5 in Mixed Use Urban Corridors: A case study of Omaha, Nebraska. *Sustainable Communities and Society*, **15**, 64-74. https://doi.org/10.1016/j.scs.2014.12.001

[8] Bernatsky, S., Smargiassi, A., Johnson, M., *et al.* (2015). Fine Particle Air Pollution, Nitrogen Dioxide, and Systemic Autoimmune Rheumatic Disease in Calgary, Alberta. *Environmental Research*, **140**, 474-478. https://doi.org/10.1016/j.envres.2015.05.007

[9] Bernatsky, S., Smargiassi, A., Barnabe, C., *et al.* (2016). Fine Particulate Air Pollution and Systemic Autoimmune Rheumatic Disease in Two Canadian Provinces. *Environmental Research*, **146**, 85-91. https://doi.org/10.1016/j.envres.2015.12.021

[10] Block, M., Elder, A., Auten, R.L. *et al.* (2012). The Outdoor Air Pollution and Brain Health Workshop. *NeuroToxicology*, **33**, 972-984. https://doi.org/10.1016/j.neuro.2012.08.014

[11] Boarnet, M.G., Houston, D., Edwards, R., *et al.* (2011) Fine Particle Concentrations on Sidewalks in Five Southern California Cities. *Atmospheric Environment*, **45**, 4025-4033. https://doi.org/10.1016/j.atmosenv.2011.04.047

[12] Brugge, D., Durant, J.L. and Rioux, C. (2007) Near-Highway Pollutants in Motor Vehicle Exhaust: A Review of Epidemiologic Evidence of Cardiac and Pulmonary Health Risks. *Environmental Health*, **6**, 23. https://doi.org/10.1186/1476-069X-6-23

[13] Brugge, D., Patton, A.P., Bob, A., *et al.* (2015) Developing Community Level Policy and Practice to Reduce Traffic-Related Air Pollution Exposure. *Environmental Jus-*

tice, **8**, 95-104. https://doi.org/10.1089/env.2015.0007

[14] Cacciottolo, M., Wang, X., Driscoll, J., *et al.* (2017) Particulate Airpollutants, APOE Alleles and Their Contributions to Cognitive Impairment in Older Women and to Amyloidogenesis in Experimental Models. *Translational Psychology*, **7**, Article ID: e1022.

[15] Caiazzo, F., Ashok, A., Waite, I.A., *et al.* (2013) Air Pollution and Early Deaths in the United States. Part I. Quantifying the Impact of Major Sectors in 2005. Atmospheric Environment, **79**, 198-208. https://doi.org/10.1016/j.atmosenv.2013.05.081

[16] CARB (2017) California's Advanced Clean Car Midterm Review, California Air Resources Board Report. http://www.arb.ca.gov/

[17] Cartier, Y., Benmarhnia, T. and Brousselle, A. (2015) Tool for Assessing Health and Equity Impacts of Interventions Modifying Air Quality in Urban Environments. *Evaluation and Program Planning*, **53**, 1-9. https://doi.org/10.1016/j.evalprogplan.2015.07.004

[18] Chen, H., Kwong, J.C., Copes, R., *et al.* (2017) Living near Major Roads and the Incidence of Dementia, Parkinson's Disease, and Multiple Sclerosis: A Population-Based Cohort Study. *Lancet*, **389**, 718-726. https://doi.org/10.1016/S0140-6736(16)32399-6

[19] CTA (2014) Electric Bus, Chicago Transit Authority. http://www.transitchicago.com/

[20] EC. (2017) Air Quality Standards, European Commission of the European Union. http://ec.europa.eu/

[21] Erickson, L.E. (2017) Reducing Greenhouse Gas Emissions and Improving Air Quality: Two Global Challenges. *Environmental Progress and Sustainable Energy*, **36**, 982-988. https://doi.org/10.1002/ep.12665

[22] Erickson, L.E. and Jennings, M. (2017) Energy, Transportation, Air Quality, Climate Change, Health Nexus: Sustainable Energy Is Good for Our Health. *AIMS Public Health*, **4**, 47-61. https://doi.org/10.3934/publichealth.2017.1.47

[23] Erickson, L.E., Robinson, J., Brase, G. and Cutsor, J., Eds. (2017) Solar Powered Charging Infrastructure for Electric Vehicles: A Sustainable Development. CRC Press, Boca Raton, Florida.

[24] Ewing, J. (2017) France Plans to End Sales of Gas and Diesel Cars by 2040. New York Times, 6 July 2017. https://www.nytimes.com/

[25] Feigin, V.L., Roth, G.A., Naghavi, M., *et al.* (2016) Global Burden of Stroke and Risk Factors in 188 Countries during 1990-2013: A Systematic Analysis for the Global Burden of Disease Study 2013. *The Lancet Neurology*, **15**, 913-924. https://doi.org/10.1016/S1474-4422(16)30073-4

[26] Filippini, T., Heck, J.E., Malagoli, C., *et al.* (2015) A Review and Meta-Analysis of Outdoor Air Pollution and Risk of Childhood Leukemia. *Journal of Environmental Science and Health C: Environmental Carcinogenesis Ecotoxicology Reviews*, **33**, 36-66. https://doi.org/10.1080/10590501.2015.1002999

[27] Fleischer, N.L., Merialdi, M., van Donkelaar, A., *et al.* (2014) Outdoor Air Pollution, Preterm Birth and Low Birth Weight: Analysis of the World Health Organization Global Survey on Maternal and Perinatal Health. *Environmental Health Perspectives*, **122**, 425-430. https://doi.org/10.1289/ehp.1306837

[28] Frangoul, A. (2017) London's Black Cabs Are About to Turn Green Thanks to a New $373 Million Electric Taxi Factory. CNBC. http://www.cnbc.com/

[29] Genc, S., Zadeoglulari, Z., Fuss, S.H. and Genc, K. (2012) The Adverse Effects of Air Pollution on the Nervous System. *Journal of Toxicology*, **2012**, Article ID: 782462.

https://doi.org/10.1155/2012/782462

[30] Ghosh, J.K.C., Heck, J.E., Cockburn, M., *et al.* (2013) Prenatal Exposure to Traffic-Related Air Pollution and Risk of early Childhood Cancers. *American Journal of Epidemiology*, **178**, 1233-1239. https://doi.org/10.1093/aje/kwt129

[31] Giles, L.V., Barn, P., Kunzli, N., Romieu, I., *et al.* (2011) From Good Intentions to Proven Interventions: Effectiveness of Actions to Reduce the Health Impacts of Air Pollution. *Environmental Health Perspectives*, **119**, 29-36. https://doi.org/10.1289/ehp.1002246

[32] Gulia, S., Nagendra, S.M.S., Khare, M. and Khanna, I. (2015) Urban Air Quality Management: A Review. *Atmospheric Pollution Research*, **6**, 286-304. https://doi.org/10.5094/APR.2015.033

[33] Ha, S., Hu, H., Roussos-Ross, D., *et al.* (2014) The Effects of Air Pollution on Adverse Birth Outcomes. *Environmental Research*, **134**, 198-204. https://doi.org/10.1016/j.envres.2014.08.002

[34] Health Effects Institute (2017) State of Global Air 2017. Health Effects Institute, Boston, MA. http://www.healtheffects.org/

[35] Heck, J.E., Wu, J., Lombardi, C., *et al.* (2013) Childhood Cancer and Traffic-Related Air Pollution Exposure in Pregnancy and Early Life. *Environmental Health Perspectives*, **121**, 1385-1391. https://doi.org/10.1289/ehp.1306761

[36] Heydarpour, P., Amini, H., Khoshkish, S., *et al.* (2014) Potential Impact of Air Pollution On Multiple Sclerosis in Tehran, Iran. *Neuroepidemiology*, **43**, 233-238. https://doi.org/10.1159/000368553

[37] Hitchcock, G. and Vedrenne, M. (2014) Cycling and Urban Air Quality. European Cyclists Federation. http://www.ecf.com/

[38] Holmes-Gen, B. and Barrett, W. (2016) Clean Air Future: Health and Climate Benefits of Zero Emission Vehicles. American Lung Association of California Report, Chicago.

[39] Hyland, J. and Donnelly, P. (2015) Air Pollution and Health: The Views of Policy Makers, Planners, Public and Private Sector on Barriers and Incentives for Change. *Journal of Transport and Health*, **2**, 120-126. https://doi.org/10.1016/j.jth.2015.03.006

[40] IARC (2012) Diesel Engine Exhaust Carcinogenic, Press Release No. 213. International Agency for Research on Cancer, World Health Organization, Lyon.

[41] Jack, O. (2016) Electric Bus Developments in Europe. BUSRide. http://busride.com/

[42] Janhall, S. (2015) Review on Urban Vegetation and Particle Air Pollution: Deposition and Dispersion. *Atmospheric Environment*, **105**, 130-137. https://doi.org/10.1016/j.atmosenv.2015.01.052

[43] Kan, H., Chen, B. and Hong, C. (2009) Health Impact of Air Pollution in China: Current Knowledge and Future Research Needs. *Environmental Health Perspectives*, **117**, A187. https://doi.org/10.1289/ehp.12737

[44] Klejnowski, K., Krasa, A., Rogula-Kozlowska, W. and Blaszczak, B. (2013) Number Size Distribution of Ambient Particles in a Typical Urban Site: The First Polish Assessment Based on Long-Term (9 Months) Measurements. *The Scientific World Journal*, **2013**, Article ID: 539568. https://doi.org/10.1155/2013/539568

[45] Krzyzanowski, M., Apte, J.S., Bonjour, S.P., *et al.* (2014) Air Pollution in the Mega-Cities. *Current Environmental Health Reports*, **1**, 185-191. https://doi.org/10.1007/s40572-014-0019-7

[46] Kuuluvainen, H., Ronkko, T., Jarvinen, A., *et al.* (2016) Lung Deposited Surface

Area Size Distribution of Particulate Matter in Different Urban Areas. *Atmospheric Environment*, **136**, 105-113. https://doi.org/10.1016/j.atmosenv.2016.04.019

[47] Lambert, F. (2017) Seattle Is Getting a Massive Fleet of All-Electric Buses, Proterra Gets the Bulk of the Order. Electrek. https://electrek.co

[48] Lelieveld, J., Evans, J.S., Fnais, M., *et al.* (2015) The Contribution of Outdoor Air Pollution to Premature Mortality on a Global Scale. *Nature*, **525**, 367-371. https://doi.org/10.1038/nature15371

[49] LeSage, J. (2017) Beijing Cab Operators Converting to Electric Taxis. Hybrid Cars. http://www.hybridcars.com/

[50] Liu, C. (2016) Electric Bus Pioneer Pulls into Wider European Market. China Daily. http://www.chinadaily.com.cn/

[51] Mani, M.S., Sandhu, S.C., Joshi, G., *et al.* (2013) India: Diagnostic Assessment of Select Environmental Challenges, Report No. 70004-IN. World Bank, South Asia Region.

[52] McDonald, Z. (2016) Electric Taxis Are on Their Way. Fleet Carma. http://www.fleetcarma.com/

[53] Mehrpour, M., Shams-Hosseini, N.S., Rezaali, S., *et al.* (2013) Effect of Air Pollutant Markers on Multiple Sclerosis Relapses. *Iranian Journal of Public Health*, **42**, 1167-1173.

[54] Meltzer, S. (2014) Six Things All Cities Should Be Doing to Reduce Urban Air Pollution Now. Urban Times. https://urbantimes.com/

[55] Morello-Frosch, R., Jesdale, B.M., Sadd, J.L. and Pastor, M. (2010) Ambient Air Pollution Exposure and Full-Term Birth Weight in California. *Environmental Health*, **9**, 44. https://doi.org/10.1186/1476-069X-9-44

[56] Morello-Frosch, R. and Shenassa, E.D. (2006) The Environmental Riskscape and Social Inequality: Implications for Explaining Maternal and Child Health Disparities. *Environmental Health Perspectives*, **114**, 1150-1153. https://doi.org/10.1289/ehp.8930

[57] Mousavi, S.E., Heydarpour, P. and Reis, J. (2017) Multiple Sclerosis and Air Pollution Exposure: Mechanisms toward Brain Autoimmunity. *Medical Hypotheses*, **100**, 23-30. https://doi.org/10.1016/j.mehy.2017.01.003

[58] OECD (2016) The Economic Consequences of Outdoor Air Pollution. OECD Publishing, Paris.

[59] O'Gorman, C., Lucas, R. and Taylor, B. (2012) Environmental Risk Factors for Multiple Sclerosis: A Review with a Focus on Molecular Mechanisms. *International Journal of Molecular Sciences*, **13**, 11718-11752. https://doi.org/10.3390/ijms130911718

[60] Oikonen, M., Laaksonen, M., Oksaranta, O., *et al.* (2003) Ambient Air Quality and Occurrence of Multiple Sclerosis Relapse. *Neuroepidemiology*, **22**, 95-99. https://doi.org/10.1159/000067108

[61] Oudin, A., Forsberg, B., Adolfsson, A.N., *et al.* (2016) Traffic-Related Air Pollution and Dementia Incidence in Northern Sweden: A Longitudinal Study. *Environmental Health Perspectives*, **124**, 306-312.

[62] Pattekar, S. (2017) Combating Pollution in Los Angeles: How One City Is Improving Air Quality. My Sidewalk. https://blog.mysidewalk.com/

[63] Peters, J., Theunis, J., Van Poppel, M., *et al.* (2013) Monitoring PM 10 and Ultrafine Particles in Urban Environments Using Mobile Measurements. *Aerosol and Air Quality Research*, **13**, 509-522.

[64] Petroff, A. (2017) These Countries Want to Ditch Gas and Diesel Cars. Money, CN News. http://money.cnn.com/

[65] Prince, M., Comas-Herrera, Knapp, M., *et al.* (2016) World Alzheimer Report 2016. Alzheimer's Disease International. http://www.alz.co.uk

[66] Raz, R., Roberts, A.L., Lyall, K., Hart, J.E., Just, A.C., Laden, F. and Weisskopf, M.G. (2015) Autism Spectrum Disorder and Particulate Matter Air Pollution before, during and after Pregnancy: A Nested Case-Control Analysis within the Nurses' Health Study II Cohort. *Environmental Health Perspectives*, **123**, 264-270.

[67] Ridlington, E. and Madsen, T. (2017) Our Health at Risk. Environment Illinois Research and Education Center. http://www.environmentillinoiscenter.org/

[68] Ritz, B., Lee, P.C., Hansen, J., *et al.* (2016) Traffic Related Air Pollution and Parkinson's Disease in Denmark: A Case Control Study. *Environmental Health Perspectives*, **124**, 351-356.

[69] Roberts, A.L., Lyall, K., Hart, J.E., Laden, F., Just, A.C., *et al.* (2013) Prenatal Air Pollutant Exposures and Autism Spectrum Disorder in the Children of Nurses Health Study II Participants. *Environmental Health Perspectives*, **121**, 978-984.

[70] Rohtaki, H. (2017) LED Boards Set Up at Five Spots to Display Air Quality Index. Indian Express, 6 June 2017. http://indianexpress.com

[71] Rowe, D.B. (2011) Green Roofs as a Means of Pollution Abatement. *Environmental Pollution*, **159**, 2100-2110. https://doi.org/10.1016/j.envpol.2010.10.029

[72] Salmond, J.A., Tadaki, M., Vardoulakis, S., *et al.* (2016) Health and Climate Related Ecosystem Services Provided by Street Trees in the Urban Environment. *Environmental Health*, **15**, 95-111. https://doi.org/10.1186/s12940-016-0103-6

[73] ScienceNordic (2015) Twelve Things That Can Protect You from Polluted Urban Air. http://sciencenordic.com/

[74] Shandas, V., Voelkel, J., Rao, M. and George, L. (2016) Integrating High Resolution Data Sets to Target Mitigation Efforts for Improving Air Quality and Public Health in Urban Neighborhoods. *International Journal of Environmental Research and Public Health*, **13**, 790. https://doi.org/10.3390/ijerph13080790

[75] Sun, G., Hazlewood, G., Bernatsky, S., *et al.* (2016) Association between Air Pollution and the Development of Rheumatic Disease: a Systematic Review. *International Journal of Rheumatology*, **2016**, Article ID: 5356307. https://doi.org/10.1155/2016/5356307

[76] TfL (2016) Mayor Unveils First Fully Electric Bus Routes for Central London, Transport for London. https://tfl.gov.uk/

[77] Trasande, L., Malecha, P. and Attina, T.M. (2016) Particulate Matter Exposure and Preterm Birth: Estimates of U.S. Attributable Burden and Economic Costs. *Environmental Health Perspectives*, **124**, 1913-1918. https://doi.org/10.1289/ehp.1510810

[78] Turpen, A. (2015) Nissan Now Has More Than 550 Electric Taxi Cabs on the Road in Europe, Torque New. http://www.torquenews.com/

[79] Underwood, E. (2017) The Polluted Brain. *Science*, **355**, 342-345. https://doi.org/10.1126/science.355.6323.342

[80] UNICEF (2016) Clear the Air for Children: The Impact of Air Pollution on Children. United Nations International Children's Emergency Fund Report, UNICEF, New York. http://www.unicef.org/

[81] United Nations (2005) Urban Air Quality Management Toolbook. United Nations Environment Program, Nairobi.

[82] U.S. EPA (2012) Air Climate and Energy: Strategic Research Action Plan 2012-2016. EPA 601/R-12/003/June 2012. http://www.epa.gov/

[83] U.S. EPA (2014) National Air Toxics Program: The Second Integrated Air Toxics Report to Congress, EPA-456/R-14-001. http://www.epa.gov/

[84] U.S. EPA (2015) Air, Climate, and Energy: Strategic Research Action Plan 2016-2019, EPA 601/K-15/005/November 2015. http://www.epa.gov/

[85] U.S. EPA (2016a) Technical Assistance Document for the Reporting of Daily Air Quality—The Air Quality Index (AQI). U.S. Environmental Protection Agency. http://www.epa.gov/

[86] U.S. EPA (2016b) EPA Awards $4.5 Million to Advance Air Monitoring Technology. U.S. Environmental Protection Agency. https://www.epa.gov/newsreleases/epa-awards-45-million-advance-air-monitoring-technology

[87] U.S. EPA (2017) National Ambient Air Quality Standards Table. U.S. Environmental Protection Agency. http://www.epa.gov/

[88] Voelker, J. (2017) London's $27 Entry Charge for Pre-2016 Diesel Cars to Start April 2019, Green Car Reports. http://www.greencarreports.com/

[89] Vojinovic, S., Savic, D., Lukic, S., *et al.* (2015) Disease Relapses in Multiple Sclerosis Can Be Influenced by Air Pollution and Climate Seasonal Conditions. *Vojnosanitetski Pregled*, **72**, 44-49. https://doi.org/10.2298/VSP140121030V

[90] Volk, H.E., Hertz-Picciotto, I., Delwiche, L., Lurmann, F. and McConnell, R. (2011) Residential Proximity to Freeways and Autism in the CHARGE Study. *Environmental Health Perspectives*, **119**, 873-877. https://doi.org/10.1289/ehp.1002835

[91] Wang, X., Ding, H., Ryan, L. and Xu, X. (1997) Association between Air Pollution and Low Birth Weight: A Community Based Study. *Environmental Health Perspectives*, **105**, 514-520. https://doi.org/10.1289/ehp.97105514

[92] Wang, L., Zhong, L., Vardoulakis, S., *et al.* (2016) Air Quality Strategies on Public Health Equity in Europe: A Systematic Review. *International Journal of Environmental Research and Public Health*, **13**, 1196. https://doi.org/10.3390/ijerph13121196

[93] WAQI (2017) Real Time Air Quality Index, Air Pollution in the World. The World Air Quality Index Project.

[94] WHO (2016) India Takes Steps to Curb Pollution. *Bulletin of the World Health Organization*, **94**, 487-488. https://doi.org/10.2471/BLT.16.020716

[95] Williams, R., Kilaru, V., Snyder, E., *et al.* (2014) Air Sensor Guidebook. U.S. Environmental Protection Agency, Washington DC.

[96] World Bank (2016) The Cost of Air Pollution: Strengthening the Economic Case for Action. World Bank Report, Washington DC.

[97] Xue, J., Li, Y., Wang, X., *et al.* (2015) Comparison of Vehicle Exhaust Particle Size Distributions Measured by SMPS and EEPS during Steady State Conditions. *Aerosol Science and Technology*, **49**, 984-996. https://doi.org/10.1080/02786826.2015.1088146

[98] Zeft, A.S., Prahalad, S., Schneider, R., *et al.* (2009) Juvenile Idiopathic Arthritis and Exposure to Fine Particulate Air Pollution. *Clinical and Experimental Rheumatology*, **27**, 877-884.

[99] Nelson, L.J. and Reyes, E.A. (2017) Metro Agrees to Buy 95 Electric Buses, in the First Step toward an Emissions-Free Fleet. Los Angeles Times, 27 July 2017.

Permissions

All chapters in this book were first published in JEP, by Scientific Research Publishing; hereby published with permission under the Creative Commons Attribution License or equivalent. Every chapter published in this book has been scrutinized by our experts. Their significance has been extensively debated. The topics covered herein carry significant findings which will fuel the growth of the discipline. They may even be implemented as practical applications or may be referred to as a beginning point for another development.

The contributors of this book come from diverse backgrounds, making this book a truly international effort. This book will bring forth new frontiers with its revolutionizing research information and detailed analysis of the nascent developments around the world.

We would like to thank all the contributing authors for lending their expertise to make the book truly unique. They have played a crucial role in the development of this book. Without their invaluable contributions this book wouldn't have been possible. They have made vital efforts to compile up to date information on the varied aspects of this subject to make this book a valuable addition to the collection of many professionals and students.

This book was conceptualized with the vision of imparting up-to-date information and advanced data in this field. To ensure the same, a matchless editorial board was set up. Every individual on the board went through rigorous rounds of assessment to prove their worth. After which they invested a large part of their time researching and compiling the most relevant data for our readers.

The editorial board has been involved in producing this book since its inception. They have spent rigorous hours researching and exploring the diverse topics which have resulted in the successful publishing of this book. They have passed on their knowledge of decades through this book. To expedite this challenging task, the publisher supported the team at every step. A small team of assistant editors was also appointed to further simplify the editing procedure and attain best results for the readers.

Apart from the editorial board, the designing team has also invested a significant amount of their time in understanding the subject and creating the most relevant covers. They scrutinized every image to scout for the most suitable representation of the subject and create an appropriate cover for the book.

The publishing team has been an ardent support to the editorial, designing and production team. Their endless efforts to recruit the best for this project, has resulted in the accomplishment of this book. They are a veteran in the field of academics and their pool of knowledge is as vast as their experience in printing. Their expertise and guidance has proved useful at every step. Their uncompromising quality standards have made this book an exceptional effort. Their encouragement from time to time has been an inspiration for everyone.

The publisher and the editorial board hope that this book will prove to be a valuable piece of knowledge for researchers, students, practitioners and scholars across the globe.

List of Contributors

Rong-Hwa Huang and Po-Han Chen
Department of Business Administration, Fu Jen Catholic University, Taiwan

Shun-Chi Yu
Department of Tourism and Hospitality, China University of Science and Technology, Taiwan

Manqing Ying, Lingjuan Wang-Li and Larry F. Stikeleather
Department of Biological and Agricultural Engineering, North Carolina State University, Raleigh, USA

Jack Edwards
Department of Mechanical and Aerospace Engineering, North Carolina State University, Raleigh, USA

Jon P. Bossley
Environmental Science Graduate Program, The Ohio State University, Columbus, OH, USA
School of Natural and Social Sciences, Mount Vernon Nazarene University, Mount Vernon, OH, USA

Peter C. Smiley Jr.
USDA Agricultural Research Service, Soil Drainage Research Unit, Columbus, OH, USA

Hiroyuki Harada, Yumi Katayama, Minori Inoue, Ryota Teranaka and Yoshiharu Mitoma
Faculty of Environmental Science, Prefectural University of Hiroshima, Hiroshima, Japan

Asmak Afriliana
Faculty of Agricultural Technology, Univerity of Jember, Jember, Indonesia

Abdullah Murrar
Faculty of Graduate Studies, Strategic Planning & Fundraising, Arab American University Jenin, Jenin, Palestine Tariff & Licensing Directorate, Water Sector Regulatory Council, Ramallah, Palestine

Ibrahim Awad
Faculty of Economics and Business Administration, Al-Quds University of Jerusalem, Jerusalem, Palestine

Abdel Fattah Hasan
Water and Environmental Studies Institute, An-Najah National University, Nablus, Palestine

Eyad Yaqob, Ihab Barghothi, Ahmad Sadaqa and Abdelrahman Tamimi
Graduate Studies Teaching Committee, Arab American University Jenin, Jenin, Palestine

Subhi Samhan
Research and Development Directorate, Palestinian Water Authority, Ramallah, Palestine

Abdelrahman Tamimi
General Directorate, Palestinian Hydrology Group, Jerusalem, Palestine

Ying Li
Department of Environmental Health, College of Public Health, East Tennessee State University, Johnson City, USA

E. O. Nwaichi
University of Massachusetts, Amherst, MA, USA

L. C. Chuku
University of Port Harcourt, Port Harcourt, Nigeria

Mahamadi Nikièma, Marius K. Somda and Désiré Traoré
Centre de Recherche en Sciences Biologiques Alimentaires et Nutritionnelles (CRSBAN), Département de Biochimie-Microbiologie, Université Ouaga I, Pr Joseph KI-ZERBO, 03 BP 7131 Ouagadougou 03, Burkina Faso

Kifouli Adéoti, Farid Baba-Moussa and Fatiou Toukourou
Laboratoire de Microbiologie et Technologies Alimentaires (LAMITA), Département de Biologie végétale, Université d'Abomey-Calavi, B.P. 526 Abomey-Calavi, Bénin

Dayéri Dianou
Centre National de la Recherche Scientifique et Technologique (CNRST), 03 BP 7192 Ouagadougou 03 Burkina Faso, Ouagadougou, Burkina Faso

Oral S. Saulters, Brijesh Krishnan, Mary Beth Harris, Hani Bashour, Laura Bellows, Evan J. Ringquist and Diane Henshel
School of Public and Environmental Affairs, Indiana University, Bloomington, IN, USA

Linda Jones
School of Psychology, Massey University, Wellington, New Zealand

Larry E. Erickson and Ronaldo G. Maghirang
Kansas State University, Manhattan, KS, USA

Wendy Griswold
University of Memphis, Memphis, TN, USA

Brian P. Urbaszewski
Respiratory Health Association, Chicago, IL, USA

Index

CPSIA information can be obtained
at www.ICGtesting.com
Printed in the USA
BVHW010856050719
552497BV00024B/6/P